IF

# OXFORD COGNITIVE SCIENCE SERIES

*General Editors*
MARTIN DAVIES, JAMES HIGGINBOTHAM, PHILIP JOHNSON-LAIRD,
CHRISTOPHER PEACOCKE, KIM PLUNKETT

Published in the series

*Concepts: Where Cognitive Science Went Wrong*
Jerry A. Fodor

*Context and Content*
Robert C. Stalnaker

*Mindreading*
Stephen Stich and Shaun Nichols

*Face and Mind: The Science of Face Perception*
Andy Young

*Reference and Consciousness*
John Campbell

*Seeing Reason: Image and Language in Learning to Think*
Keith Stenning

*Ways of Seeing: The Scope and Limits of Visual Cognition*
Pierre Jacob and Marc Jeannerod

# IF

## Jonathan St B T Evans
*University of Plymouth*


## David E Over
*University of Sunderland*

OXFORD
UNIVERSITY PRESS

# OXFORD
UNIVERSITY PRESS

Great Clarendon Street, Oxford OX2 6DP

Oxford University Press is a department of the University of Oxford.
It furthers the University's objective of excellence in research, scholarship,
and education by publishing worldwide in

Oxford  New York

Auckland  Bangkok  Buenos Aires  Cape Town  Chennai
Dar es Salaam  Delhi  Hong Kong  Istanbul  Karachi  Kolkata
Kuala Lumpur  Madrid  Melbourne  Mexico City  Mumbai  Nairobi
São Paulo  Shanghai  Taipei  Tokyo  Toronto

Oxford is a registered trade mark of Oxford University Press
in the UK and in certain other countries

Published in the United States
by Oxford University Press Inc., New York

© Jonathan Evans and David Over 2004

A catalogue record for this title is available from the British Library

ISBN 0–19–852512–5 (Hbk)

ISBN 0–19–852513–3 (Pbk)

10 9 8 7 6 5 4 3 2 1

Typeset by Newgen Imaging Systems (P) Ltd., Chennai, India
Printed in Great Britan
on acid-free paper by
Biddles Ltd, King's Lynn

*To the memory of Peter Wason (1924–2003)*

# Preface

We have been interested in the word 'if' throughout our academic careers. JE started his work on conditionals with his PhD (1972) under the supervision of the psychologist, Peter Wason, who almost alone created, from nothing, the experimental study of conditionals. DO's interest in 'if' also dates from his PhD (1974). He was equally lucky in his supervisor, Dorothy Edgington, who has done much to advance the philosophy of conditionals. The philosophy and psychology of conditionals have been conducted in relative isolation from one another. One objective of the present book is to try to begin a process of integrating these two fields. Another is to argue for the integration of the psychology of conditional reasoning with research on probability, utility, and closeness judgements in the psychology of judgement and decision making.

What we present in this book is essentially a psychological approach to 'if'. We do not try to answer deep normative or, more broadly, philosophical questions about conditionals. However, our approach has been strongly influenced by philosophical logic. Our belief in the importance of the philosophy to the psychology was one of the major stimulants for the research that led to the writing of this book. Another was our conviction that leading psychological theories of conditionals were seriously inadequate. We develop our theoretical view of conditionals from the more general psychological framework of a previous collaboration, *Rationality and Reasoning* (Evans and Over, 1996a). In particular, we draw upon our more general theory of hypothetical thinking and the dual process theory of reasoning.

There are a number of people whose help we are happy to acknowledge. We have worked closely with Simon Handley in both experimental and theoretical research that has led us to the view of conditionals that is presented in this book. We have been most fortunate in the collaborators we have had in our research on conditionals: John Clibbens, Aidan Feeney, David Green, Dinos Hadjichristidis, Ken Manktelow, Steve Newstead, Steven Sloman, Valerie Thompson, and the late Rosemary Stevenson. Other people who have especially helped us over the years in discussions of issues in the book are Ruth Byrne, Nick Chater, Kit Fine, Vittorio Girotto, Wilfred Hodges, Colin Howson, Dick Jeffrey, Phil Johnson-Laird, Danny Kahneman, Hans Kamp, Paolo Legrenzi, Jonathan Lowe, Henry Markovits, Mike Oaksford, Guy Politzer, Dan Sperber, Keith Stanovich, and Rich West. The literature that has influenced us should be clear from our references. But we are particularly grateful to Dorothy Edgington and Klaus Oberauer, not only for discussion, but for their critical reading of an earlier draft of this book. We have not always listened to our collaborators, critics, and readers, and the remaining mistakes in the book are ours.

Direct support for the writing of this book came from several sources. Our most recent experimental work on conditionals was supported by a grant from the Economic and Social Research Council (R000239074). Most of the writing was done while the authors were on study leave from their universities. JE received a generous period of sabbatical leave from the University of Plymouth during which he benefited from the extraordinary tolerance and patience of his wife, Jane. DO had a vital Research Leave grant from the Arts and Humanities Research Board (AN10017/APN16040) and was also given generous leave by the University of Sunderland. He benefited from the support of his colleagues in Psychology at Sunderland, and from the facilities he was given as a Visiting Professor in the Psychology Departments of the Universities of Durham and Plymouth. He was grateful for Dartmoor National Park, Hadrian's Wall Path National Trail, and Shenandoah National Park, where he was able, only too briefly, to escape 'if'. He had the greatest possible support from his wife, Marilyn, and daughter, Harriet.

JE, DO
*March 2004*

# Contents

# 1 'If' and the problem of hypothetical thought

'If' is one of the most important and interesting words in the language. It is used to express hypothetical thought, which is an essential part of human reasoning and decision making. This type of thought occurs wherever there is uncertainty, and uncertainty is everywhere in human affairs. A teacher is not certain that a student will pass an exam, but she thinks that he will pass *if* he starts to work hard. The police do not know beyond reasonable doubt that a suspect is guilty, but they believe that he is guilty *if* his fingerprints are on the murder weapon. A chemist is unsure what some liquid is, but knows that it is an acid *if* it turns litmus paper red. 'If' helps us to imagine and analyse possible states of the world, and to think about how things might be now or in the future, or how they might have been in the past.

Sentences of the form, 'if p then q', are termed conditional statements, or just conditionals for short. In this book, we are concerned with ordinary conditionals of this form in natural language. Such conditionals are common in ordinary or everyday contexts, and even in much scientific discourse, as we have just illustrated. Philosophers have long studied the meaning, usage, and inferential properties of these conditionals. More recently, linguists and psychologists have joined in, and the literature in psychology alone on conditionals is daunting. In spite of all this scholarly effort, the word 'if' is still far from well understood, and its study raises deep and complex issues yet to be resolved. Philo of Megara scandalized his fellow philosophers in the fourth century BC by claiming that 'if p then q' is equivalent to 'not-p or q' (Kneale and Kneale, 1962, chapter 3; Sanford, 1989, chapter 1). One of his examples was apparently, 'If it is day then I am conversing', which he thought was equivalent to, 'Either it is not day or I am conversing.' A conditional for which this equivalence holds is sometimes said to be a Philonian conditional, but is usually called a *material conditional* or a *truth functional conditional*. There are still some philosophers and psychologists who claim that ordinary conditionals are material conditionals, and they still scandalize most of their peers (see Chapters 2 and 4).

There has, however, been progress in our understanding of conditionals since ancient times. Conditionals, as we have said, help us to cope with uncertainty, but conditionals themselves are more or less uncertain, or to put it another way, they are more or less probable. The teacher knows that her student sometimes panics in exams, and so she has only moderate confidence that, if he starts to work hard, then he will pass. The police are more confident that, if the suspect's fingerprints are on the murder weapon, then he is guilty, but they can conceive of ways in which an innocent man's fingerprints could have got there. The chemist is the most confident of all and judges it to be extremely probable that, if the liquid turns litmus paper red, then it is an acid. As far as we know, the ancient philosophers did not ask questions about the probability of conditionals, and they could hardly have answered the questions if they had asked them. We now have probability theory to give us substantial help us with these questions. More than that, we have in psychology the experimental and statistical tools to investigate the probability judgements people actually make about conditionals. A major goal of this book is to report the experimental investigations of these probability judgements, and to draw out the consequences of what has been discovered about them (see especially Chapters 8 and 9). One experimental finding can be mentioned at this point. People do not judge the probability of 'if p then q' to be equal to the probability of 'not-p or q' (Evans *et al.*, 2003a; Over and Evans, 2003). This result alone makes it very hard to see how 'if p then q' in natural language can be equivalent to 'not-p or q'.

Conditional or hypothetical thinking is a universal feature of human cognition and so of course there is nothing special about the English rendering of the word 'if'. For example, although most psychological studies of conditional reasoning (for reviews see Evans *et al.*, 1993; Manktelow, 1999) have been run in the English language, there are quite a number using other European languages, especially French and Italian, and some using Asian languages such as Cantonese or Japanese. The phenomena that psychologists report are essentially unaffected by these various linguistic contexts. Hence, the word 'if' that is investigated in this book should be taken to represent any device used to communicate conditional statements, regardless of the particular language in which it is implemented.

Conditional statements are ubiquitous in everyday discourse. We will illustrate some of the variety of conditionals in this chapter. An extremely common use of 'if' is to form indicative conditionals. These conditionals can themselves be used in a wide variety of ways, but in one common use they can be justified by causal relations:

1.1   If you turn the key, the engine will start
1.2   If a metal rod is heated, then it will expand
1.3   The clock will run if you wind it daily

There is a fundamental question about whether 'causal' conditionals semantically state the existence of causal relations, or whether they are merely justified by causal relations. Psychologists generally use 'causal conditional' in way that does not beg this question, and we will follow this practice here. We will discuss experimental data relevant to the question in Chapters 8 and 9. There are also indicative conditionals that are justified by temporal relations, e.g. 'If the first train goes to Plymouth, then the next one goes to Exeter.' Causal conditionals can be supported by scientific laws, as in 1.2, but are often used by ordinary people to give advice, as in 1.1 and 1.3.

Another important type of conditional statement is referred to as a *counterfactual*, which is commonly used when there is some pragmatic suggestion or presupposition that its antecedent and consequent are false. Clear examples are about an action that was once, but is no longer, possible:

1.4   If you had avoided the motorway, you would have got there quicker
1.5   If I had worked harder, I would have passed my exams

Counterfactuals have received considerable attention in the philosophical literature and more recently in the psychological literature (see Chapter 7). People often think about what might have been, and use counterfactuals to express this. Such thoughts can have a considerable effect on people's immediate emotions and future decision making. For example, consider a student who has the thought expressed by 1.5 and who reflects further that he could easily have worked harder. Perhaps he usually works hard, but failed his most recent exams because he became overconfident and did not do enough work. He is likely to feel bitter regret when he has his counterfactual thought, and to resolve to work hard in the future. Here we see that the study of reasoning with conditionals should make close contact with psychological research on decision making and even, in this case, with the psychology of emotions such as regret.

Another type of conditional is the *deontic* conditional. Indicative and counterfactual conditionals are used to try to describe the way the world is, might be, or might have been. We would use the indicative conditional 1.1 to describe what will start the car. We would use counterfactual 1.4 to describe the way to avoid heavy traffic. Deontic conditionals, in contrast, are used, not to describe, but rather to try to guide or regulate behaviour. These conditionals express rules and regulations and often contain the deontic modals, the deontic 'must', 'should', and 'may', as in:

1.6 If the traffic light is red then you must stop
1.7 If you have an exam coming up, then you should work hard
1.8 If you buy a licence, then you may fish in this river

The use of deontic conditionals is yet another example of why there should be a very close relation between the psychology of conditional thought and the psychology of decision making. People continually refer to rules and regulations—prudential, social, and moral—when they are trying to decide what to do. In fact, trying to make a decision is just of matter of thinking about what one *should* or *may* do, which often means thinking about one should or may do *if* some condition holds (Over *et al.*, 2004).

An interesting class of conditionals is used to influence the listener's behaviour by offering advice or inducements. Such conditionals are frequently used to offer advice or inducements with speech acts that can be differentiated as conditional promises, tips, threats, and warnings as in the following examples:

1.9 If you avoid the motorway then you will get there quicker (tip)
1.10 If you clean my car then you can borrow it tonight (promise)
1.11 If you smoke cigarettes then your health will suffer (warning)
1.12 If you arrive late again, I will fire you (threat)

Tips are partly distinguished from promises, and warnings from threats, on the grounds of whether or not the speaker has direct control over the consequent event. This distinction that has been shown psychologically to affect the inferences that people draw from such speech acts (Newstead *et al.*, 1997; Evans and Twyman-Musgrove, 1998). All these types of conditionals are uttered by a speaker to a listener in a communicative context that takes us into the realm of pragmatics (discussed below in this chapter). We cannot, for example, begin to understand what a conditional (as opposed to unconditional) promise means, without examining the motives of the person who utters it. Studying these speech acts forces us to relate reasoning and decision making very closely to each other. The antecedent of one of these conditionals refers to an action that the listener can decide to take. For that reason, it would hardly make sense for the listener to ask for the probability that the antecedent is true. Unfortunately, psychological research on conditionals has yet to make adequate contact with psychological research on decision making. One goal that we have in this book is to make a start on overcoming this limitation.

Are these conditionals indicative or deontic? They seem to have some characteristics of each. Inducements—tips and warnings—often relate to causal relations as in 1.11, which would seem to render them indicative. One could debate whether such assertions are true, in the sense of accurately describing relations in the world. However, they are not merely about states of affairs, but uttered in a context that is intended to regulate or modify someone's behaviour. Inducements—threats and promises—have a more clearly deontic nature. Consider the threat of 1.12 above. This seems superficially indicative because you could read it as a causal relation (being late will cause you to be fired) or ask to what extent it is true that being late will result in your being fired. On the other hand, there is a strong pragmatic (see below) implication that 'If you want to keep your job you must not be late', which expresses a deontic rule. One way to view these types of conditionals then is that they are indicative in form but carry pragmatic implications of a deontic nature.

Another topic for us to discuss will be conditionals with negations in them, as these have been extensively studied in the psychological literature (see especially, Chapters 3 and 5). Many 'if' statements are modified by the word 'not', in either the antecedent or consequent or both. To understand the nature of such conditionals, we need to give some brief attention to the concept of negation itself. The word 'not', like the word 'if', has both a logical meaning and a linguistic and pragmatic usage, which goes well beyond it. The logical meaning of 'not' is that of reversing truth and falsity. As observed by the famous psychologist Peter Wason (1972), however, negative statements tend not to be used pragmatically in everyday language to assert information, but make 'plausible denials' by denying presuppositions. Thus we can assert happily that a dolphin is not a fish, but it would be pragmatically anomalous to assert that a horse is not a fish.

Our main interest in 'not' is that it combines itself effortlessly and naturally with 'if'. Negated conditionals are perfectly natural in all the categories of conditional statements discussed above. Here are some examples to prove the point:

1.13   If you go out drinking you won't get up in the morning
1.14   If you don't clean my car then you can't borrow it tonight
1.15   If I had not driven so fast, then I would not have got here in time
1.16   If you don't have an invitation then you must be a member to come in here

Note that these examples include conditionals with negations in the antecedent, consequent, or both, and are of the advice or inducement forms (1.13, 1.14) as well as counterfactual (1.15) and deontic (1.16) forms.

'Not' is such a close linguistic friend to 'if' that it seems that negation is built into our facility for hypothetical thinking. In imagining another state of the world, we can find it equally easy to imagine something to be false that is actually true, or imagine something true that is actually false. With future indicatives, we can base our thinking on a positive possibility (what if we take up salsa dancing) or a negative one (what if we do not go to the meeting). Similarly, counterfactual thoughts can be positive (what if we had gone for a long walk) or negative (what if we had not had too much to drink).

# Theoretical approach

In this book, we will begin to develop a psychological theory of conditionals and hypothetical thought distinct from existing psychological accounts in the literature. Although our aim is a psychological theory, we will make reference to the literature in philosophical logic on conditionals where this is relevant to our interests, as it often is. We have been most influenced by those philosophical logicians who have argued that the ordinary conditional of natural language is not a material conditional, meaning 'not-p or q', but is rather to be evaluated using the so-called Ramsey test (see Chapter 2). This 'test' is a type of hypothetical thought that evaluates the probability of 'if p then q' as the conditional probability of q given p, and not as the probability of 'not-p or q'. Persuasive philosophical arguments have been given for the Ramsey test, and we have experimental evidence in support of it (see Chapter 8). However, the Ramsey test is a high-level description of many processes in hypothetical thinking that psychologists will have to explain in detail.

The Ramsey test alone is not sufficient in itself to account for how people understand and use conditionals. We must also take account of pragmatic processes that introduce additional knowledge and belief cued by the context in which a conditional is used. And we need dual process theory to explain how people can reason either pragmatically or deductively according to their general cognitive ability and the way in which they are instructed. The Ramsey test, pragmatics, and dual process theory are all relevant to the detailed interpretation of psychological experiments on conditional reasoning. We will discuss the background to the Ramsey test in Chapter 2. We introduce briefly here our view of the importance of pragmatics and dual process theory.

## Pragmatics and relevance

So far we have done little more than convey by example the wide variation in linguistic usage of 'if'. To go further, we should shift our attention to the pragmatic level. Pragmatics concerns the way in which prior knowledge and belief influences communication between people. Pragmatics goes well beyond a purely linguistic analysis of the syntax and semantics of sentences. We may communicate by linguistic or non-linguistic means and—very commonly—by a combination of both. Much of our communication is implicit, resting upon shared beliefs and knowledge, presupposition of motives and the like.

For example, suppose you observe a woman rush into her house and hear her husband say, 'It is on the table.' This communication, though verbal, has no clear meaning that could be derived by linguistic analysis. You would need to know that the woman had phoned her husband 10 minutes earlier to say that she had forgotten her passport and that she was turning around on her drive to the airport. In understanding this sentence, she would assume that her husband had made a statement relevant to her goal and that he had found her passport in the place it was usually kept and placed it on the table to save her time. Language was used here, but need not have been. For instance, he might have simply pointed to the table.

Even where people have a normal conversation with completed sentences, communication rests on pragmatics (see Sperber and Wilson, 1995, for numerous examples).

Natural language does not operate like a computer language with fully explicit syntax and semantics that are sufficient to achieve communication. Knowledge, belief, and supposition by the speaker and listener about each other's states of mind are almost always found in successful communication. We are not going to get very far in our quest to understand the use of 'if' in everyday discourse without consideration of the pragmatic level. Much pragmatic inference, most of it rapid and implicit, takes place in communication. Pragmatic inference is very different from explicit logical inference or deduction. In the latter, we restricted ourselves to explicitly stated premises and draw only conclusions that are necessary given the assumed truth of those premises. By contrast, in a pragmatic inference, we can make use of any relevant premise that the context allows, take account of how believable the premise is, and draw from it a conclusion that is probable to some degree.

Let us reconsider one of our previous examples of a conditional promise:

1.10   If you clean my car then you can borrow it tonight

Using your own pragmatic inference system, you can probably draw a number of inferences about this statement before we give you any context for it at all, such as:

- The speaker has some form of power relationship with the listener in which he/she has relative wealthy and authority. Quite probably this is an utterance from a parent to a young adult child.

- The speaker has the goal of having their car cleaned and does not wish to do this work themselves. Quite probably the listener is responsible for the messy state of the car when borrowing it previously.

- The speaker believes that the listener has the goal of borrowing the car.

- The promise is conditional because the speaker wants to influence the behaviour of the listener, specifically to make them clean the car.

- If the car is not cleaned, then it probably won't be lent. However, you might not have high confidence in this inference, because you know that parents can be soft and children persuasive and that statements such as 'I haven't got time now, but I will do it in the morning' have a reasonable success rate in such situations.

These inferences have everything to do with pragmatics and little to do with logic. We can make all of these inferences as outside observers, and we could probably make much more confident inferences if we had some background knowledge about the speaker and listener.

Much pragmatics theorizing derives from the work of Grice (1975) and his notion of a pragmatic *implicature*. Grice proposed several rules for conversation that are followed by speaker and listener. One of Grice's maxims that is highly relevant to conditionals is that you should not say less than you mean. For example, you would not say 'Some football matches are exciting' if you believed that 'All football matches are exciting', even though the statement is logically consistent with your beliefs. There is some evidence in the psychological literature that people draw Gricean rather than logical inferences with such statements (Newstead and Griggs, 1983). As applied to conditional promises, threats, and so on, the point is that you should not make a statement conditional, if you could assert the consequent unconditionally. Hence,

1.10   If you clean my car then you can borrow it tonight

is by pragmatic implication biconditional. We would probably not state a condition for borrowing the car if we were going to lend it anyway. Hence, if the car is not cleaned, you do not get to borrow it. The problem, as already noted, is that although the speaker uses the conditional to try to make the listener clean the car, the listener may know from previous experience that the speaker is a soft touch. Thus the pragmatic inferences of the listener may go beyond the rules of conversation and the intended meaning of the speaker. We shall have more to say about Grice in Chapter 6.

Sperber and colleagues (Sperber and Wilson, 1995; Sperber *et al.*, 1995) recently updated their relevance theory of pragmatics, originally published in 1986. What was previously known as the principle of relevance is now the *communicative* principle of relevance, which has been augmented by a second, cognitive principle of relevance in the new account (see Chapter 5). The communicative principle of relevance is essentially that all communications convey a guarantee of their own relevance. Sperber and Wilson believe this principle can effectively replace the separate Gricean maxims.

The addition of the cognitive principle of relevance—whose elaboration we will not go into for the present—is most welcome as it adds what is to our mind an essential psychological component to the theory. This echoes our own theoretical approach to the understanding of 'if' in this book. Not only must be go beyond a logical and linguistic treatment of 'if', but we must go beyond pragmatics as well. Ultimately, the understanding of 'if' must relate to a more general psychological theory of hypothetical thinking. We briefly review the background to this in the following section.

## Dual process theory and hypothetical thinking

Evans (1984, 1989) presented what is known as the heuristic-analytic theory of reasoning, primarily in order to account for a variety of observed cognitive biases in reasoning and judgement tasks conducted in the psychological laboratory. The theory distinguished between *heuristic* processes responsible for forming selective representations of problem content and *analytic* processes that reason with such representations. The term 'heuristic' in this theory refers to processes that operate pragmatically at a preconscious level, determining automatically what gets represented as 'relevant'. These representations include both selective features of problem content and also relevant associated knowledge retrieved from long-term memory. Analytic processes refer to more or less effective procedures for generating inferences and decisions from such information. Evans (1989) argued that many errors and biases arose not from failures in analytic reasoning so much as from the nature of the representation of problem information. If logically relevant information is excluded during problem representation or logically irrelevant information included, then this will normally lead to an observed bias.

The heuristic-analytic theory was applied, among other things, to the explanation of biases observed in psychological experiments on conditional reasoning, as we shall see in later chapters. However, the psychological distinction between two types of processing was elaborated and developed in the dual process theory of Evans and Over (1996a), influenced by Reber's (1993) dual process account of implicit and explicit learning. In this account, there are two distinct cognitive systems, later labelled as System 1 and

System 2 by Stanovich (1999b). (For a related distinction, see Sloman, 1996.) System 1, variously described as implicit, unconscious, or pragmatic, retrieves and applies knowledge in a rapid and automatic manner. It appears to operate through associative neural networks and is independent of individual differences in working memory capacity or measured general intelligence. It is hypothesized to have evolved early and to form the universal basis for animal cognition.

System 2, by contrast, is hypothesized to have evolved late and uniquely in human beings. Evans and Over (1996a) suggest that is linked to other uniquely human facilities of language and reflective consciousness. This system, which may be labelled explicit, conscious, or analytic, operates through verbal working memory, is relatively slow and sequential in nature and does correlate in its efficiency with measured general intelligence (see Stanovich, 1999; Stanovich and West, 2000). Whereas System 1 processes, which provide such facilities as language comprehension and pattern recognition appear to have been optimized by evolution to universally high levels of functioning, System 2 processes reflect the high heritability of cognitive capacity, as measured by IQ. (Heritable characteristics vary between individuals according to the genes passed on by parents.) In particular, the ability to suppress pragmatic influences and to reason abstractly is higher in people of relatively high cognitive ability. These people have an increased ability, not only to infer logical solutions to conditional and other reasoning problems, but also to find normative solutions to a range of decision and judgement problems as well. However, System 2 is a general human faculty, and there is some logical and other abstract thinking ability in all human beings (Stanovich, 1999b).

The dual process account of reasoning holds that both System 1 and System 2 cognition are present in all human beings (Evans, 2003; see also Fodor, 1983; Kahneman and Frederick, 2002). The defining characteristics of System 2 are its slow, sequential nature, its dependence upon working memory capacity (related to the IQ link), and its independence from associative and pragmatic processes. This, of course, is the system that provides the 'analytic' reasoning in the Evans (1989) theory. But what exactly does System 2 do, and what unique advantages does it provide for human beings?

In a previous writing on this topic (Evans and Over, 1996) we examined, more or less empirically, a variety of cases where System 2 thinking was demonstrably present in reasoning and judgement tasks. We came to the conclusions that all were manifestations of one or another kind of hypothetical thinking. That is to say of thought that required representation of possibilities. The relevant functions, as measured in the cognitive tasks, include deductive reasoning, hypothesis testing, and consequential decision making. If we take decision making as an example, we can see that humans, in common with all higher animals, can make decisions based upon past experience. When we choose a restaurant franchise on the basis of past satisfaction with meals eaten, for example, our behaviour is being shaped in Skinnerian fashion by past reinforcement history in just the same ways as has been demonstrated in numerous animal experiments. This is universal System 1 decision making.

Unlike other animals, however, humans *can* make decisions in a different, System 2 manner. We can project the future consequences of our actions and imagine the possible outcomes that will result from them. We can then choose on the basis of the relative probability and utility of these possibilities, just as decision theory says we should. Of course, the psychological literature on decision making is littered with evidence of biases and failures to choose consequentially in decision tasks (Shafir *et al.*, 1993; Baron, 1994), just as the psychological literature on deductive reasoning is marked by widespread evidence of logical error. This is testimony to the dominance of System 1 thinking in most people, most of the time. However, we are not the same as other animals. The uniquely human System 2 allows us to make inferences that are unavailable to other animals and ultimately to acquire scientific knowledge and technologies to help us deal with an uncertain world.

The ability of other animals to cope with changing and uncertain environments is limited by the extremely slow process of evolution by natural selection and by the still relatively slow process of conditioning. Natural selection has given animals conditional instincts that will deal with uncertainty to a limited extent. A blackbird will give an alarm call *if* it sees a fox. Conditioning and simple learning can also give animals conditional abilities. The fox will learn to run *if* it hears the hunter's horn. However, these are System 1 implicit conditionals. Human beings have the System 2 ability to engage in hypothetical thought and, on this basis, to formulate explicit conditionals, performing explicit inferences from them and using them in communication. System 2 has itself certainly evolved by natural selection, although exactly how it has done so is still unclear (Over and Evans, 2000; Over, 2003). System 2 has evolved by natural selection on top of simple System 1 learning, just as increasing ability for System 1 learning in animals as intelligent as foxes as evolved on top of simpler instincts.

In fact, we see the link between System 2 explicit conditionals and conditional probability as ultimately grounded in System 1 learning. There must always be some conditionals, of the form 'if p then q', that people consider probable to the extent that they have learned that q type events follow p type events. However, people are not restricted to such cases, but can also use, for example, mental models of complex causal relationships to make probability judgements about conditionals. We begin by judging it probable that, if we go to a departmental meeting, we will get a headache, because we have learned, using System 1, to associate headaches with such meetings. Expressing this experience in a conditional statement, e.g. 'If I go to a departmental meeting, I get a headache', however, would involve System 2. If we had purely System 1 intelligence, we might avoid the meeting, much as a rat avoids a location where it received an electric shock. However, System 2 is required for hypothetical reasoning, such as when we say to a friend, 'I am not going to the departmental meeting, because they always give me a headache'. We would use System 2 to construct and reason about more refined mental causal models in which the headache depends on who chairs the meeting, who speaks at it, how long it goes on for, and other factors. We can then assert more discriminating causal conditionals with probability judgements that are generated theoretically and not just on the basis of past experience.

We have developed our ideas about hypothetical thinking in recent years. In particular, we propose three general principles (Evans *et al.*, 2003b) to describe it. According to the *singularity* principle, we generally only consider one possibility, hypothesis or mental model at a time. By the *relevance* principle, the possibility or model considered is that which is most relevant in the current context, and by default this will be the most probable or plausible possibility. Finally, the *satisficing* principle states than rather than trying optimize choice among alternatives (a practical impossibility in a complex and effectively unbounded world), we maintain the current model provided that it satisfies our constraints and goals to a sufficient degree. However, we will abandon and replace the current model when evidence is encountered that shows it to be unsatisfactory. This theory implies a strong role for System 1 thinking in determining the relevance of possibilities. However, representation of possibilities in hypothetical thinking must be explicit, as part of this representation is our knowledge that they are hypothetical. Without this stipulation, we would have no way to distinguish reality from supposition. We also suppose that explicit System 2 reasoning is involved in the evaluation of models needed to implement the satisficing principle. The theory of 'if' to be developed in this book takes place against this background.

## Approach and form of the book

As already indicated, we believe that a psychological account of 'if' must be based on a theory of hypothetical thinking. However, we most emphatically do not believe that psychologists have a monopoly of wisdom on this topic. Indeed, we regard the relative inattention to the writings of linguistics, logicians, and philosophers on the topic of conditionals in the psychological literature to be a major weakness. Similarly, we believe that philosophical logicians would benefit also from being better informed about the extensive findings of experimental psychologists on this topic. Hence we plan an interdisciplinary approach in which we consider the contributions of authors from these differing backgrounds. In particular, we wish to attempt to integrate philosophical argument and psychological evidence with regard to all the major issues concerning conditional statements.

In Chapter 2, we examine the major logical issues concerning indicative conditional statements, as identified by philosophical logicians. We show why philosophers largely reject the view that 'if p then q' is equivalent to 'not-p or q', and discuss some of the problems and complications that have arisen from trying to understand the logic of the ordinary conditional. In Chapter 3, we then examine the basic empirical findings on indicative conditionals when participants are asked to make inferences or judgements about them in the psychological laboratory and compare these findings with the analyses of logicians. In Chapter 4, we examine the treatment of conditionals within the psychological theories of mental logic and mental models.

In Chapter 5, we summarize and discuss the massive investigation of conditional thinking that has been conducted by psychologists using the Wason four card selection task (Wason, 1966). We share the views of various contemporary authors that the selection task has probably been overworked as a paradigm for studying human deductive reasoning. However, the focus of this book on not on deduction *per se*, but on all aspects of the usage and understanding of conditional statements and the associated hypothetical thinking that they engender. We believe that, correctly interpreted, the research on the Wason task has much to contribute to this enterprise. The selection task has been one of the main vehicles for the investigation of content and context effects in recent years, the discussion of which leads us into a more general examination of conditionals in context, including causal conditionals, in Chapter 6. We then look in some detail at the important topic of counterfactual conditionals (Chapter 7), examining both philosophical and psychological contributions.

As already indicated, our own theory of conditionals is developed from a simple psychological proposal by the philosopher Ramsey, whose work is discussed in Chapter 2. This ultimately leads us to a suppositional and probabilistic account of conditionals that we believe to be consistent with the bulk of the psychological evidence that is reviewed in this book. In an uncertain world, hypothetical thinking must deal not just with possibilities, but with probabilities. Hypothetical thinking serves our personal goals to which the probabilities of antecedent and consequent events, and especially the conditional probability of the latter given the former, may be critically relevant. Probabilistic representation of conditional statements has been discussed in detail by philosophers but has been investigated experimentally by psychologists only in recent years. Contributions from both disciplines are discussed in Chapter 8. Finally, in Chapter 9, we expand and develop our theory of conditionals, attempting in the process to resolve the issues and account for the data presented in the book as a whole.

# 2 The logic of indicative conditionals

In Chapter 1, we saw that conditional statements have a number of different purposes and applications. In this chapter, we will consider only indicative conditionals. Linguists would find it too simple or crude to say that these conditionals are in the indicative mood, but saying that gives some idea of the topic.

Philosophical logicians and cognitive psychologists who have studied the indicative conditional in natural language have been deeply influenced by the great advances in formal logic late in the nineteenth century and early in the twentieth century. From the very beginning, these works included a formal conditional with attractively simple logical properties (Frege, 1970, originally published in 1879). This is the material conditional that we referred to briefly in Chapter 1. The material conditional is *truth functional*—it is often called the truth functional conditional—and is logically equivalent to 'not-p or q'. The technical term 'truth functional' here means that the *truth value*, i.e. the truth or falsity, of the material conditional is fully determined by, and is indeed a strict mathematical function of, the truth or falsity of its component propositions. Where 'if p then q' is a material conditional, it is false when we have p true and q false, and otherwise it is true. Exactly the same can be said, of course, about 'not-p or q', and we should really identify the material conditional with 'not-p or q'. Whether indicative conditionals in natural language are material and therefore truth functional is a major issue to be addressed in this book.

We will follow Edgington (2001, 2003) in distinguishing, very broadly, between three different accounts of the indicative conditionals in natural language, termed simply T1, T2, and T3. These are not specific theories in themselves, but rather cover three different families of theories. T1 states that indicative conditionals in natural language are material, truth functional conditionals. If T1 were correct, then the semantic truth conditionals and the valid inference rules of the indicative conditional in natural language would be the same as those of the truth functional conditional. T2 agrees with T1 that ordinary conditionals are propositional, in the sense of being true or false at every possibility, but in T2 these conditionals are not truth functional. Finally, T3 conditionals, termed *suppositional* by Edgington (2003), might be called 'semi-propositional'— they are true or false only when the antecedent holds. We will start with a detailed examination of T1 and present our argument that this cannot be the correct theory of the ordinary conditional. We then consider theories of types T2 and T3.

## Extensional semantics and T1

The most perspicuous way to represent the truth functional conditional is to use the *extensional semantics* of a truth table (Wittgenstein, 1961, first published in 1921).

**Table 2.1.** Truth values for a truth functional conditional *if p then q*

| Possible states of affairs | Truth values of p and of q | Material conditional: 'if p then q' |
|---|---|---|
| TT | T  T | T |
| TF | T  F | F |
| FT | F  T | T |
| FF | F  F | T |

Table 2.1 is a truth table. It gives the semantics of the truth functional conditional by displaying its truth conditions. Table 2.1 provides the truth conditions of the material conditional, which according to T1 applies also to the ordinary indicative conditional. In the table, T and F are the truth values, representing truth and falsity. TT is the possible state of affairs in which both p and q are true, TF is the possible state of affairs in which p is true and q is false, FT is the possible state of affairs in which p is false and q is true, and FF is the possible state of affairs in which both p and q are false. These four cases exhaust the logical possibilities that can be derived from all the combinations of truth and falsity for p and q. Table 2.1 clearly shows that a material conditional has the following truth conditions. It has the truth value T for TT, FT, and FF and the truth value F for TF.

Consider an ordinary indicative conditional in natural language:

2.1   If it rains (r), then the plants will die (d)

As we will use this example a lot, we will refer to 2.1 as RD. For this example, the logically possible states of affairs will be:

TT   r & d            It rains and the plants die
TF   r & not-d        It rains and the plants do not die
FT   not-r & d        It does not rain and the plants die
FF   not-r & not-d    It does not rain and the plants do not die

If RD is a material conditional, as held by T1, then RD is true in three possibilities, where 'r & d', 'not-r & d', and 'not-r & not-d' hold, and false in case that 'r & not-d' holds.

The material or truth functional conditional can also be called the extensional conditional. Its formal logic is represented in *truth functional*, or *extensional, propositional logic*. The precise specification of this elementary logic was the necessary condition for all the great advances in logic that followed it (Frege, 1970). The compound propositions in extensional propositional logic are generated from component propositions using the truth functional connectives: negation, conjunction, disjunction, and the truth functional conditional. For example, the compound proposition:

(p & q) or (not-p & not-q)

has '(p & q)' and '(not-p & not-q)' as its immediate components and is logically equivalent to a conjunction of truth functional conditionals:

(if p then q) & (if q then p)

The above is the truth functional biconditional, which can also be written:

p if and only if q

For example, suppose we state:

2.2   You are pregnant if and only if the test will be positive

Assuming that 2.2 is a material biconditional, as in T1, we would have to conclude that 2.2 is logical equivalent to holding that one of the following extensional conjunctions is true:

You are pregnant and the test will be positive
You are not pregnant and the test will not be positive

A compound proposition in this logic is always a truth function of its components. This relation between the compound and its components determines the truth conditions and so the very meaning of the compound.

The truth functional or material conditional is sometimes called material implication (Whitehead and Russell, 1962, originally published in 1910). In this book, we will prefer the terms 'truth functional' or 'material conditional' and will avoid the term 'material implication'. There is a certain danger in using 'material implication', 'materially implies', or just 'implies' to refer to or to express the truth functional conditional. The problem is that this might be read as a causal link or some other kind of real world connection, whereas no such relationship is in fact indicated. For example, if 2.1 is a truth functional conditional, 2.1 (RD) asserts no more nor less than: 'It does not rain or the plants will die.' Clearly, this disjunction does not state that 'it rains' implies in any intuitive sense that 'the plants will die', nor does it state a causal or any other intuitive relation between rain and the death of the plants. And if the disjunction does not make these statements, then RD, assumed to be a material conditional, cannot do so either. If RD, as a natural language conditional, is stronger than this disjunction, then it cannot in fact be a truth functional conditional. This is very important, as we shall see.

The material conditional 'if p then q', being logically equivalent to 'not-p or q', is the weakest major premise that can be added to p as a minor premise to infer q as a conclusion in valid reasoning. Inferring q as a conclusion from the premises p and 'if p then q' is the classically valid inference form called Modus Ponens (MP). An ordinary example of this form of inference is:

If it rains then the plants will die
It is raining
Therefore, the plants will die

The above instance of MP is essentially the same as inferring q from p and 'not-p or q', if RD is a material conditional. Given that either it will not rain or the plants will die, and we are told that it is raining, this eliminates the first disjunct. As a result, the second—plants dying—must necessarily follow.

Another classically valid inference is Modus Tollens (MT):

If it rains then the plants will die
The plants will not die
Therefore, it will not rain

This instance of MT is essentially the same as inferring not-p from not-q and 'not-p or q', if RD is a material conditional. In this case we eliminate the second disjunct, so the first must be true—it will not rain. But again, a basic question for us will be whether RD is a material conditional or whether it is something stronger in ordinary language.

A logically *valid* inference, by definition, cannot possibly have true premises and a false conclusion (all other combinations of truth and falsity are possible). A valid inference logically guarantees that its conclusion is true given that its premises are true: its conclusion is logically necessary given its premises. However, a valid inference makes no promises about the truth or falsity of its premises. One or more of the premises of a valid inference can be false, and then there is no guarantee at all about the truth of its conclusion. We can also state the definition of validity in this way:

> An inference is valid if and only if there is no logically possible state of affairs in which its premises are true and its conclusion false

Both MP and MT can for the material conditional be proven valid by means of truth tables (Table 2.2). In a truth table analysis, we enumerate all the logical possibilities by supposing that the component propositions are either true or false. In this case there are just two propositions, p and q, and so just four possibilities to consider. Because the conditional is material, it is only false when p is true and q is false, and otherwise it is true. We see that there is no line in either the MP or MT truth tables in which the conclusion is false when both premises are true, so the arguments are valid. Note that a *sound* argument can be defined as a valid argument with true premises. MP will be sound when the state of the world is TT, and MT will be sound when the state of the world is FF. When instances of MP and MT are sound, their conclusions must, of course, be true.

It is correspondingly easy to prove inferences invalid in truth functional logic. Consider the classical fallacies of conditional reasoning. There is Affirmation of

**Table 2.2.** Truth table proofs for Modus Ponens and Modus Tollens arguments

(a) Modus Ponens

| Possible states of affairs p q | Major premise 'If p then q' | Minor premise 'p' | Conclusion 'q' |
|---|---|---|---|
| T T | T | T | T |
| T F | F | T | F |
| F T | T | F | T |
| F F | T | F | F |

(b) Modus Tollens

| Possible states of affairs p q | Major premise 'If p then q' | Minor premise 'not-q' | Conclusion 'not-p' |
|---|---|---|---|
| T T | T | F | F |
| T F | F | T | F |
| F T | T | F | T |
| F F | T | T | T |

the Consequent (AC):

> If it rains then the plants will die
> The plants will die
> Therefore, it will rain

and there is Denial of the Antecedent (DA):

> If it rains then the plants will die
> It will not rain
> Therefore, then plants will not die

We could write out the truth tables for these two inferences as we did for the valid inferences, as shown in Table 2.2. If we did so, we would discover a counterexample to both inferences, that is a case in which the premises are true and the conclusion false. This obtains for AC when the state of the world is FT: the material conditional is true, the minor premise, q, is true, but the conclusion, p, is false. FT also provides a counter-example to the DA inference. In fact, these inferences would only be valid if the truth table for the conditional defined FT as false. Such a truth table represents the material biconditional, 'p if and only if q', for which all four inferences are consequently valid.

Psychologists have discovered that people often endorse the conditional fallacies, AC and DA (see Chapter 3). In response to this finding, some authors have tried to preserve the claim that the ordinary conditional is truth functional by suggesting that it is implicitly read as a material biconditional in some contexts. This is sometimes referred to as the 'chameleon' theory of conditionals (Braine, 1978) because the conditional is supposed to change its colour from (material) conditional to (material) biconditional as the context requires. For example, the view is that many people would interpret, 'If you are pregnant then the test will be positive', as the (supposedly material) biconditional 2.2. We do not think that the chameleon theory can defend the claim that the ordinary conditional is a material conditional. However, we will bear this theory in mind in reviewing the psychological evidence later in this book.

Ordinary conditionals in natural language logically imply the material conditional, even though, as we hold, the reverse is not true. Our example, conditional RD, whether truth functional or not, certainly rules out the possibility of r true and d false. In other words to accept this conditional is also to reject the possibility that it is raining and the plants will not die. According to T1, it follows further that inferences that are valid for the material conditional are also valid for the natural language indicative conditional. However, as we will argue in the next section, against T1, there are valid inferences for the material conditional that are *invalid* for the ordinary indicative conditional.

## Inference rules

Logically possible states of affairs and the truth tables that represent them are the basis of the logical semantics, or logical model theory, of truth functional logic. The clearest way to give logical inference rules for truth functional logic, in its proof theory, is to use a natural deduction system (Gentzen, 1969, originally published 1934), so called because it was thought that people reason using its rules in ordinary life. In this system, the propositional connectives have introduction and elimination inference rules that can be proved valid by reference to truth tables and their logically possible states of affairs. Each connective has an introduction rule, for 'introducing' it into an argument, and an elimination rule for 'eliminating' it from an argument.

We will consider as an example the introduction and elimination rules for 'and'. The rule of and-introduction tells that we may infer 'p & q' when we have already established (or assumed) that both p and q hold. For instance, suppose we know from personal experience that a friend of ours, Linda, is a feminist. Linda then tells us that she has become a banker. The rule of and-introduction allows us to infer as the next step that Linda is a feminist and a banker. This conclusion must be true given that the premises are true (that we do know that Linda is a feminist and Linda has told us the truth about her new job). The rule of and-elimination allows us to infer q, or equally p, when we have already established (or assumed) that 'p & q' holds. For example, if we already know that Linda is a feminist and a banker, we may validly infer that she is a banker. It is impossible for the conjunctive premise to be true and the conclusion false.

These inferences may seem too trivial to be useful, but trivial inferences can be strung together to derive highly informative and even extremely surprising results. The introduction and elimination rules for the truth functional connectives precisely reflect the fact that these connectives are truth functional. As we have already pointed out, the truth values of these connectives are strict mathematical functions of the binary values, truth and falsity. Many real-world processes and states can be described by using equivalent binary values and truth functions of these. Computers are the best example of an application of this fact. Computers can represent binary values, such as truth and falsity, by using 1 and 0 (and at the deepest level by high voltage and low voltage). They do all their informative and useful work for us by means of so-called 'logical gates', which simply embody truth functional connectives. For example, there is a logic gate, a so-called AND gate, that precisely embodies the truth function for 'and'. Of course, the introduction and elimination rules for 'and' precisely reflect this truth function as well. Computers can use their logic gates to perform many more inferences much faster than we can by means of natural deduction rules. However, our limited processing speed does not always prevent our inferences from being equally useful.

People are also prone to mistakes and biases in what should be 'trivial' reasoning. Inferring q from 'p & q' may seem trivial, but people sometimes overlook the validity of this inference. Tversky and Kahneman (1983) showed that, after being given a description of Linda as a supposedly typical feminist, people tend to judge it more probable that Linda is a feminist *and* a banker than that Linda is a banker. Yet this is obviously a biased probability judgement in light of the validity of and-elimination. It is logically impossible for Linda to be a feminist and a banker, but not a banker, and so the former cannot be more probable than the latter. People do recognize the relation between validity and probability in problems with a transparent logical structure (Sloman and Over, 2003). An advantage of explicitly using a natural deduction system is that it can help us to see logical relationships very clearly.

We are perfectly happy that and-elimination, as given in truth functional logic, is a valid inference rule for the ordinary use of 'and', although not all philosophers and psychologists appear happy with this (Strawson, 1952, and see Chapter 4). However, we must now consider the introduction and elimination rules for the truth functional conditional. The rule of if-elimination is easy to state: it is simply MP. The rule of truth functional if-introduction is a little more difficult to state. Suppose that we have used the rules of the natural deduction system to infer q from a set of assumptions S plus the extra supposition p. The rule of if-introduction, for the truth functional 'if', allows us to infer 'if p then q' from the set S alone.

As an example of if-introduction in truth functional logic, consider the following two assumptions:

2.3 If Linda is a feminist and a banker, then she is a banker
2.4 If Linda is a banker, then she will vote for the Conservative Party

Now let us make the extra supposition that Linda is feminist and a banker. From assumption 2.3, we may validly infer that Linda is a banker (by MP or if-elimination). Then from assumption 2.4, we may infer that Linda will vote for the Conservative Party (again by MP or if-elimination). This allows us to conclude by truth functional if-introduction:

2.5 If Linda is a feminist and a banker, then she will vote for the Conservative Party

The conclusion 2.5 no longer depends upon the extra supposition, but only on the original two assumptions, 2.3 and 2.4.

We hope, by our example, to have already sown a doubt that truth functional if-introduction is a valid rule for the ordinary indicative conditional in natural language. It seems to us easy to imagine a state of affairs in which 2.3 and 2.4 are true, but 2.5 is false. Assumption 2.3 is a logical truth on any account and so must be true in all examples. Imagine then that Linda is a non-feminist banker who will vote for the Conservative Party. That makes 2.4 true, but we do not believe that 2.5 will be necessarily true in this possibility. We will return to this example to explain it more fully, and describe a much better rule of if-introduction for ordinary conditionals, later in this chapter. But we can be clear now that truth functional if-introduction is invalid, in our view, for ordinary conditionals.

The invalidity of truth functional if-introduction for ordinary conditionals can be seen as one of the so-called paradoxes of material implication. These are not well named, as there is no paradox in having a formal system with a weak, truth functional conditional in it logically equivalent to 'not-p or q'. The 'paradoxes' are the absurd consequences of claiming that *ordinary* conditionals are truth functional. There are two basic paradoxes in this sense:

P1 Given not-p, it follows that if p then q
P2 Given q, it follows that if p then q

The validity of these inference rules for the material conditional is easy to see by reference to the truth table, Table 2.1. There is no possible state of affairs in which the premise of either inference is true and the conclusion false. It is also possible to prove their validity using truth functional if-introduction (and some other inference rules of natural deduction).

The absurdity comes from claiming that P1 and P2 are valid for ordinary conditionals in natural language. Consider applying P1 to RD as an example of an ordinary conditional:

It will not rain
Therefore, if it rains, then the plants will die

Suppose that there has been a drought and the plants are starting to die as a result. In this case, it seems absurd that RD should be true (by P1) merely because it will not, in fact, rain. Equally, we will not assert RD merely because we believe strongly that the drought will continue. Yet if RD is a truth functional conditional, RD would be true, and so we would apparently be justified in asserting RD on the mere basis that the drought will continue.

Now let us apply P2 to RD:

The plants will die
Therefore, if it rains, then the plants will die

We can continue imagining the same possible state of affairs about the drought to feel the absurdity of this inference as well. So what if the drought will, in fact, continue and the plants will die as a consequence? How could that fact alone make RD true and be justification for asserting RD? Surely we have described a possible state of affairs, when there is a drought and the plants need water, in which the premises of both inferences are true but RD is false. That such a possible state of affairs is so easy to imagine is what makes it appear paradoxical to claim that these inferences are valid for the ordinary indicative conditional in natural language.

Accepting P1 and P2 as valid for the ordinary conditional would also force many absurd probability judgements on us. Suppose a woman who knows that she is almost certainly not pregnant goes to see her doctor and is told, 'If you are pregnant then the test will be positive.' Suppose as well that the woman has noticed that the doctor has picked up the wrong test. It is a test for diabetes, and she is almost certain that she does not have this disease. If the doctor's statement is a material conditional, or the material biconditional 2.2, then the woman should have a great deal of confidence in it. She should conclude (by one instance of P1) that the (diabetes) test will be positive if she is pregnant, and (by another instance of P1) that she is pregnant if the (diabetes) test is positive. The premises of the supposedly valid inferences (the two instances of P1) have very high subjective probability for her, and so if she is to be unbiased, she should conclude that the conclusions also have at least this level of probability. But of course, in real life, the woman would have no confidence in the doctor's conditional, whether she interpreted it as just a conditional or as a biconditional. And she would be right. The doctor's conditional is grossly unworthy of belief, but that means that it cannot be a material conditional or material biconditional.

Note that a conditional is truth functional if P1 and P2 are valid for it. To see why this must be so, consider the four logical possibilities (as in Table 2.1) for RD: if it rains then the plants will die. In both the TT and FT cases, it is true that the plants will die (d), so RD is true if P2 is valid. Similarly, in the FF case, it is false that it will rain (r), so RD follows if P1 is valid (and 'not' is truth functional). Finally, in the TF case, RD is false on any account. We have already noted that P1 and P2 are valid for the truth functional conditional. Hence a conditional is truth functional if and only if P1 and P2 are valid for it.

Some philosophers and psychologists have claimed, for various reasons, that natural language indicative conditionals are truth functional. They have proposed, in Edgington's terminology, theories in the T1 family (Jackson, 1987; Grice, 1989; Johnson-Laird and Byrne, 1991, 2002). They have had to argue that the paradoxical inference rules only appear problematic for ordinary indicative conditionals, but are really valid for them. Other philosophers and psychologists have argued against T1 that the paradoxical rules are invalid for natural language indicative conditionals, and that these conditionals are not truth functional (Stalnaker, 1968; Rips and Marcus, 1977; Evans *et al.*, 2003a; Over and Evans, 2003). Philosophers and psychologists in the former group need to give a plausible psychological explanation of why the inferences appear paradoxical for the ordinary indicative conditional, though they are supposedly valid for it.

There is a wide range of psychological evidence against T1, the view that the ordinary indicative conditional is material, which we will begin to cover in Chapter 3. In the remainder of this chapter, we examine the views of philosophers and psychologists who have rejected T1 and the material conditional interpretation. They have proposed other approaches, in either the T2 or T3 families, to understanding the ordinary conditional of natural language. There are differences among these theorists, but they at least agree that P1 and P2 are invalid for ordinary indicative conditionals, which cannot then be truth functional. Ramsey(1990, originally published 1931)took the crucial first step towards this view, and what he said is central to our account of ordinary conditionals. We discuss the basic idea of the Ramsey test prior to an attempt to distinguish T2 from T3 theories of the conditional.

## The Ramsey test

Ramsey (1931/1990) stated what came to be known as his 'test' for ordinary conditionals in natural language. Ramsey's comments were very brief and sketchy, and are open to a number of interpretations. We take Ramsey to have stated a 'test' for the degree of confidence one should have in the ordinary conditional by relating it to conditional probability. We will use the term 'probability' throughout this book in the subjectivist or Bayesian sense, referring to the degree of belief that a person has in an uncertain possibility. In these terms, Ramsey suggested that people who are arguing about 'if p then q' in natural language are trying to make subjective probability judgements about q given p, '. . . fixing their degrees of belief in q given p'. They do this by '. . . adding p hypothetically to their stock of knowledge and arguing on that basis about q . . . ' (Ramsey, 1931/1990, p. 247).

To clarify this test, let us apply it to RD. To do this, we must know, or assume, more about the circumstances in which RD is considered. The truth and meaning of ordinary conditionals is always relative to the context in which they appear and the background beliefs that appear to be relevant. Suppose first that the ground is waterlogged after a flood. To judge RD following Ramsey, we would hypothetically add the supposition that there will be rain to our knowledge about the state of the ground. In that case, our confidence in d given r would be high, as more rain would probably kill the plants under these conditions, and we would assert RD with confidence. Suppose second a different circumstance. There has been a drought and the plants are suffering for that reason. Following Ramsey, we would again hypothetically add to our knowledge the supposition that there will be rain, but this time that knowledge would be about the drought. Now our confidence in d given r would be very low. In this case, rain would probably help and not harm the plants, and we would not assert RD. So our degree of belief in the conditional statement, 'if p then q', is the same as our belief in q given p. From a Bayesian viewpoint, this means formally that P(if p then q) = P(q|p).

We get different results from the Ramsey test in the two circumstances we have described, even if these are, at a higher level, both not-r possibilities, i.e. cases in which it will not rain. Perhaps it will not rain after the flood, and perhaps there will not be rain to end the drought. If RD is a material conditional, then it must be true in both these not-r possibilities. But that is paradoxical: RD is only worthy of belief and assertion in the not-r possibility in which the ground is waterlogged. Only here does it make sense to assert that the plants will die *if* it rains (even though we believe it will not). The Ramsey test does not generate this paradox. On the basis of the Ramsey test, and simple knowledge about plants, we will correctly believe and assert RD when the ground is waterlogged, but not after a drought. The probability of d given r would be high in the former case but low in the latter.

Stalnaker (1968) extended the Ramsey test to cases where, before we can make use of Ramsey's original suggestion, we must modify our assumptions or beliefs. Suppose that we know Linda to be a non-feminist banker who votes for the Conservative Party, but we hear someone assert 2.5: 'if Linda is a feminist and a banker, then she will vote for the Conservative Party'. Stalnaker would say that we should not evaluate this assertion by hypothetically adding the antecedent of 2.5, that Linda is a feminist and a banker, to our 'stock of knowledge'. Doing that would conflict, or be downright inconsistent, with what we know or think we know about Linda. Stalnaker suggests instead that we make the minimal changes to our assumptions or beliefs necessary to remain consistent after we add the antecedent to them hypothetically. In our example, that would mean setting aside hypothetically our conviction that Linda is not a feminist.

Notice the importance of making suppositions in a way that aims to minimize the changes necessary to achieve consistency. It would be psychologically difficult to make extensive hypothetical changes to our beliefs. Even worse than that, gratuitous or unnecessarily changes to our beliefs about Linda would be of less relevance to an evaluation of 2.5 for a discussion about Linda. Thus we suppose Linda to be a feminist banker rather than a non-feminist banker, but otherwise leave what we believe about her personality unchanged. After we do this, what degree of belief should we have in the consequent of 2.5, that Linda will vote for the Conservative Party? We might answer this question simply by recalling that we know several feminists with personalities more or less like Linda's, and that none of them ever vote for the Conservative Party. This could be enough for us to assign a very low conditional probability to the consequent of 2.5 given its antecedent. That in turn will, if Ramsey and Stalnaker are right, give us very low confidence in 2.5 as a conditional.

Consider again, the two related conditionals:

2.3 If Linda is a feminist and a banker, then she is a banker

2.4 If Linda is a banker, then she will vote for the Conservative Party

We could also evaluate 2.4 by recalling that we know quite a number of bankers like Linda who do vote for the Conservative Party. That would give us high confidence in 2.4, if Ramsey and Stalnaker are right. We would have absolute confidence in 2.3 because 2.3 is a logical truth. Thus overall, we would coherently assign very high probabilities to 2.3 and 2.4 but very low probability to 2.5. That means that inferring 2.5 from 2.3 and 2.4—which we derived earlier—must be invalid for the ordinary conditional, even though it is valid for the material conditional. This example thus constitutes evidence against T1. The way we have used reference classes in a Ramsey test in our example shows why the truth functional version of this rule can produce absurd results. In our example, this rule would, in effect, force us to infer that most feminist bankers, with personalities like Linda, will be Conservative Party voters. This false conclusion would effectively follow, in an obvious fallacy, from two true assumptions: that all feminist bankers are bankers, and that most bankers, with personalities like Linda, will vote for the Conservative Party.

Johnson-Laird and Byrne (2002) object to Stalnaker's version of the Ramsey test because, they charge, trying to make hypothetical consistent changes in our beliefs, whether minimal or not, can be a considerable psychological problem. However, this is a problem that must be faced one way or another in order to maintain a belief system and is not simply a problem that arises in understanding conditionals. *If* Johnson-Laird and Byrne, for example, encounter some strong and reliable evidence against their mental model theory of reasoning, they will have to adjust some of their beliefs—however, cherished—to accommodate this new evidence. It seems odd to us to argue that we cannot make such cognitive changes when processing a conditional statement, as we must perforce do something similar whenever our beliefs are contradicted by experience. Stalnaker is simply proposing, plausibly, that we go through a similar process hypothetically that we must go through anyway when unexpected events force us to change our beliefs.

Stalnaker's extension of the Ramsey test avoids the conditional paradoxes when we are highly confident of not-p or of q. Considering RD, suppose we firmly believe that, in the actual state of affairs, it will not rain (in the relevantly near future) and the plants will die because of a drought. This is the FT possibility, in which not-r and d both hold. But if we apply Stalnaker's Ramsey test to RD, we will not infer that RD must be true given this actual state of affairs in the world. What we will do is to imagine that, contrary to our beliefs, a world in which it does rain, thus ending the drought. The truth of the conditional depends upon the plausibility of the consequent in this hypothetical state of the world and in this state it would probably be false. This extended Ramsey test directs our thought to a consistent possible state of affairs in which r holds, and away from our firm belief that not-r and d actually hold. The extended test tells us to suspend this belief in the state of affairs in which not-r and d hold.

The Ramsey test is a psychological proposal in our view, but in stating it, Ramsey and Stalnaker had clear philosophical interests and goals. They did not formulate the test as a psychological hypothesis that could be studied in experiments. We will say more later in the book about own attempt to develop a version of the Ramsey test as a psychological account of how people make judgements about conditionals (see also Evans *et al.*, 2003a; Over and Evans, 2003). We will be brief ourselves at this point about our approach. We do not believe that a full account of indicative conditionals in natural language can be given by an extensional account that represents logical possibilities and nothing else, because this leads back to T1 and the truth functional conditional. (This applies also to the mental model theory of the conditional, which we discuss in Chapter 4.) The Ramsey test is intuitively right, to begin with, in stating that people hypothetically focus on the general p possibilities, and not the general not-p possibilities, when making a judgement about 'if p then q' in some context. We will present in due course substantial psychological evidence to support this claim.

What mental processes actually implement the Ramsey test? Philosophers have said almost nothing about this empirical question, but that is understandable, as it is more one for psychology rather than philosophy. What is much less understandable is that philosophers have said little about how the Ramsey test *should* be implemented. They have not even said very much about how 'hypothetical' or 'suppositional' thought should differ from belief (but see Levi, 1996). There is not much in the direction of a normative theory of the Ramsey test in the philosophical theory, which is a pity, as a normative theory can often give pointers to psychology.

Philosophers do talk about the Ramsey test as if it were always totally in System 2, as we would put it in our dual process theory (see Chapter 1). They describe it as if it were a process that is consciously controlled, explicit, and sequential in nature. For example, they would think of an evaluation of 2.4 as always beginning with an explicit supposition that Linda is a banker, and then a continuing with sequential inference steps. In these explicit steps, there would be minimal changes to retain consistency, and an attempt to reach a conclusion about how likely it was, under the supposition, that Linda voted for the Conservative Party. Now as we see the Ramsey test, it does require System 2, as it represents possibilities, specifically antecedent possibilities. However, we have suggested (Evans *et al.*, 2003a; Over and Evans, 2003) thinking of the Ramsey test as a process that can sometimes largely take place in System 1, grounded in heuristics that are automatic and implicit. In our proposal, people begin to evaluate a conditional by forming a mental model of the antecedent and the consequent, and comparing that, either explicitly or implicitly, with a mental model of the antecedent and the negation of the consequent. The evaluation of 2.4 might begin in System 1, with a heuristic that automatically focuses our attention on a mental model in which Linda is a banker and votes for the Conservative Party. This model might fit our stereotype of a banker better than the mental model in which Linda is a banker and does not vote that way. By means of the representativeness heuristic (Tversky and Kahneman, 1983), this System 1 process could make us highly confident that Linda votes for the Conservative Party given that she is a banker. Sometimes the Ramsey test can be much more of a System 2 process, in which one mental model is judged more probable than another because of a series of explicit inferences from an explicit supposition. But it does not have to be like that.

People will sometimes primarily rely on System 1 processes and sometimes much more on System 2 processes (although we hold that the latter processes must always depend on the former to some extent). People's System 1 processes can sometimes enable them to record sample relative frequencies in simple cases, i.e. the number of times q type events that have followed p type events. In Chapter 1, we used the example of the frequency with which we have headaches after departmental meetings. This experience automatically gives us a firm conviction that we will get a headache given that we go to one of these meetings. Sometimes people can make straightforward relative frequency judgements, using System 2, when they are given explicit frequency information in word problems (Evans *et al.*, 2003b; Oberauer and Wilhelm, 2003; also see Chapter 8). We also pointed out in Chapter 1 that people can build mental causal models to make discriminating conditional probability judgements, and this calls on System 2 processing as well. Psychologists who have studied judgement and decision making, inductive inference, causal reasoning, and other mental processes have suggested many ways in which people can make conditional probability judgements. Some of these processes, of course, result in biases. For example, people can have an availability bias (Tversky and Kahneman, 1973) by trusting unrepresentative examples that easily come to mind, and they can be overconfident in their conditional probability judgements (for a recent overview of research on judgemental biases see Gilovich *et al.*, 2002). Trying to answer the question of how the Ramsey test is implemented is a problem. However, we should be clear that there is no one answer to the question. There are rather many answers that will refer to many psychological processes (see Chapter 9).

## Intensional semantics and T2

The difference between the T1, T2, and T3 families is illustrated in Table 2.3. T1 has the truth table for the material conditional as discussed above. T2 accounts propose that the conditional is propositional, being always true or false, but that you need more knowledge than just the truth table case to decide whether or not it is true. In T3, to which we will return, conditionals only have a truth value in possible states of the world in the antecedent holds.

Stalnaker (1968, 1975) presented a semantic analysis, or model theory, of the ordinary conditional derived from his extension of the Ramsey test. His analysis is the best example of a theory in the T2 family. Some of these conditionals are true, but others are false, when their antecedents are false. Stalnaker developed his version of T2 by asking what the hypothetical state of mind in the Ramsey test represents. He answered that it is an attempt to represent a possible state of the world that differs minimally from the way the world is taken to be, but in which the antecedent of the conditional is true. Stalnaker's analysis is usually expressed using the notion of relative 'closeness' between possibilities, and we can do this in the terminology we prefer in this book. Suppose we are given a Stalnaker conditional, 'if p then q', and a possible state of the world, FT, in which p is false and q is true. Is the Stalnaker conditional true or false in

**Table 2.3.** Truth table representations for conditionals of type T1, T2, and T3 (based on Edgington, 2003)

| Possible states of affairs p q | T1 Material | T2 Stalnaker | T3 Suppositional |
| --- | --- | --- | --- |
| T T | T | T | T |
| T F | F | F | F |
| F T | T | T/F | |
| F F | T | T/F | |

FT? It is true if q is true in the 'closest' possibility to FT in which p is true, and otherwise it is false. In other words, this conditional is true in FT if and only if TT is a closer possibility to FT than TF is.

Suppose that RD is a Stalnaker conditional, and assume we are in a possible state of affairs FT during a drought. It will not rain and the plants will die. But consider the closest state of affairs in which it is raining. That will be a specific TF state of affairs in which rain brings the drought to an end and the plants do not die. In that case, RD will be false by Stalnaker's analysis of the ordinary conditional. In another example state of affairs of type FT, where the ground is waterlogged, TT will be the closest state of affairs in which it is raining, and RD will be true by Stalnaker's analysis.

Notice how the Stalnaker conditional is not truth functional. The truth functional conditional is automatically true in any state of affairs of general type FT. It does not matter whether the ground is waterlogged or whether there has been a drought. No matter what the more specific state of affairs is like, the material conditional is true. All that matters is that the antecedent of the conditional is false and its consequent is true, and that makes a material conditional true. In contrast, the Stalnaker conditional is true in some specific FT states and false in others. Similarly, the material conditional is automatically true in any state of affairs of general type FF, but the Stalnaker conditional is true in some specific FF states and false in others. Hence the Stalnaker conditional is not truth functional, and its logic is not given in standard extensional propositional logic. The Stalnaker conditional is not an extensional connective but rather intensional connective. Its logic is an intensional, or modal, logic, and its formal semantics, or model theory, is given in terms of possible states of affairs, or 'possible worlds' (Stalnaker and Thomason, 1970).

Modal logic deals with necessity and possibility. To assert necessarily p is to say that p is true in all possible states of affairs, but to say that possibly p is to assert that p is true in some possible states of affairs. Compare the effect of an extensional operator such as 'not' and a modal operator like 'necessarily'. The difference is that 'not-p', but not 'necessarily-p', has a truth value that can be decided simply by considering the current state of the world. Consider the assertion that our horse will not win the race. This is false if our horse wins and is otherwise true. Now consider the assertion that our horse will necessarily win the race. This is false if our horse will lose. But even if our horse will, as it happens, turn out to win, that does not mean that it will *necessarily* win, that is that it will win in all possibilities.

To be clearer, we need to distinguish further between logical necessity and causal necessity. Logical necessity is a concept of limited value, as the only things that are logically necessary are logical truths, e.g. it is logically necessary that q follows from 'p & q'. It is not logically necessary that our horse will win the race. There are obviously logically possible states in which it loses. However, we would probably be presupposing the weaker sense of causal necessity if we asserted, in an ordinary conversation, that our horse will necessarily win the race. In that case, we would justify our assertion by reference to a significant factor that would cause our horse to win. For example, we might have nobbled the other horses with a drug that will cause them to run slow. In technical terms, we would say that our horse wins, not in all possibilities full stop, but in all causally relevant possibilities where that causal factor is present. Causal necessity is an important concept for the evaluation of conditionals using the Ramsey test.

We often base our conditional probability judgements on what we know or believe about causal factors. Statements about causal necessity are not truth functional, as they depend, not on the truth values of their component propositions alone, but also on the existence of the causal factors.

Suppose we assert, 'If we nobble our horse then it will win.' Imagine that we are honest (as we are) and our horse is slow: this puts us in the FF state of affairs. Our conditional assertion must be true if it is a truth functional conditional. But it is absurd to hold that our assertion is true. It is much more satisfying to see our assertion as a Stalnaker conditional, in which case it is false. Our horse loses in the closest possible state of affairs in which we nobble it. We know this on the basis of our causal knowledge. It would be a different story supposing we had asserted, 'If we nobble the other horses then our horse will win.' Our horse wins in the closest possible state of affairs in which we nobble the other horses. Nobbling causes horses to lose races: these horses lose in all causally relevant states of affairs.

It may help to summarize briefly the difference between Stalnaker's T2 and the T1 family. In T1, an ordinary indicative conditional, 'if p then q', has purely extensional truth conditions, being automatically true in the possible states TT, FT, and FF, and false in TF. In Stalnaker's T2, the conditional has intensional truth conditions. The Stalnaker conditional is true in TT, as the closest possibility to TT in which p is true is TT itself and q is true in TT. The Stalnaker conditional is false in TF, as the closest possibility to TF in which p is true is TF itself and q is false in TF. So far the Stalnaker conditional may look like the truth functional conditional, but the two differ with respect to FT and FF, where the antecedent p is false. A Stalnaker conditional is true in FT if TT is the closest possibility to FT in which p is true, and it is false in FT if TF is the closest possibility to FT in which p is true. Similarly, a Stalnaker conditional is true in FF if TT is the closest possibility to FF in which p is true, and it is false in FF if TF is the closest possibility to FF in which p is true. In short, all truth functional conditionals are true in FT and in FF, but some Stalnaker conditionals are false in FT or false in FF, and some Stalnaker conditionals are true in FT or true in FF. The Stalnaker conditional is thus not truth functional.

## Psychological implications of the Stalnaker T2 conditional

It might seem that philosophical talk of possibilities is distant from psychological reality. Of course, we would leave psychology behind if we thought of a representation of a possibility as specifying, in full detail, a possible state of the world. We cannot fully describe a possible state of the world by using TT, TF, FT, or FF. These possibilities concern only two propositions. With just one more proposition, we would have to consider TTT, TTF, TFT, and so on, making eight possibilities in all. A combinatorial explosion clearly threatens. However, we can argue that there are mental models of possibilities at a high level of generality. And obviously, every possibility, no matter how fully described, must fall under one of the high-level general possibilities: TT, TF, FT, or FF.

It might also seem that there is little psychological reality in talking about the relative closeness of possibilities to each other. However, psychologists in judgement and decision making have found that people do have beliefs, which even affect emotions such as regret, about the relative closeness of possibilities (Kahneman and Miller, 1986; Roese, 1997, 2004; Teigen, 1998, 2004). One implication of all this research is that there is a tight interconnection between people's probability judgements and their closeness judgements. Sometimes closeness beliefs are based on the relative frequencies of events. For example, suppose that we almost always take an umbrella out with us when there is a forecast of rain, but that we get careless on one occasion, forget our umbrella, and get soaked. Here is a case in which we will tend to think of the possibility of our remembering to take the umbrella as a relatively 'close', and we will tend to feel regret. In contrast, suppose second that we hate umbrellas, as more trouble than they are worth, and almost never go out with one. In this case, we will think of the possibility in which we go out with an umbrella as a relatively 'distant' and will not feel regret about being without one when we got soaked.

Consider how these points could be applied to a conditional such as RD. Let us suppose that we quite firmly believe that we are in a general not-r state of affairs, i.e. that it is not going to rain. Which possibility, TT or TF will we judge to be closer to us? That will depend on whether there has been a drought or the ground is waterlogged. Our background beliefs are relevant, which is why the Stalnaker conditional cannot be truth functional. If there is a drought, we will judge the world in which rain is helpful to be closest and hence reject RD as false. If the ground is waterlogged, then we will judge the world in which the plants die to be closest and hence decide that RD is true.

The Ramsey test and the Stalnaker conditional look very attractive to us as the logical and philosophical grounding for a psychological theory of indicative conditionals. But unfortunately, the Ramsey test and the Stalnaker conditional do not fit together as neatly as Stalnaker (1968) supposed. We have already shown that an obvious interpretation of the Ramsey test equates the believability of the conditional statement with conditional degree of belief in q given p. That implies, more formally, that P(if p then q) = P(q|p) (we discuss psychological evidence for this hypothesis in detail in Chapter 8). However, the probability of a Stalnaker (T2) conditional *cannot* in general equal the conditional probability P(q|p). This is one of the consequences of a famous proof by Lewis (1976) about conditionals and conditional probability (see also Hájek, 1989; Edgington, 1995; Bennett, 2003), the basis of which can be quite simply explained. As we have illustrated above, the Stalnaker conditional is sometimes true in a not-p, a FT or FF, possibility. This happens when a TT state is the nearest possibility, to FT or FF, in which p holds. The difficulty is that P(q|p) is independent of the probabilities of FT and FF. But the Stalnaker conditional will sometimes be true in FT or FF, and so must acquire some of its probability from FT or FF. The necessary result is that the probability of the Stalnaker conditional cannot, in general, be identical with the conditional probability.

Let us illustrate the point. Suppose that we assert the following conditional as we are about to carry an antique Spode teapot, made of the finest bone china, across the flagstone floor of our kitchen:

2.6   If the teapot is dropped (d), then it will break (b)

As we start across the floor, we will think of the future as having three possible branches: TT, TF, and FF. We can virtually rule out the FT possibility, in which the teapot is not dropped but breaks anyway, thinking of that as having close to 0 probability. The TT possibility, in which the teapot is dropped and breaks, has a much higher relative probability for us than the TF possibility, in which the teapot is dropped and does not break. This relative probability judgement could come from our ill experience of dropping bone china in the past on our flagstone floor, or it could come from a causal model of what happens when fine bone china impacts flagstones. However, we intend to be careful, and we assign the highest probability to the FF possibility, in which the teapot is not dropped and does not break. We might summarize these probability judgements as follows:

| | | |
|----|------------------------------------------------|------|
| TT | The teapot is dropped and it breaks | 0.14 |
| TF | The teapot is dropped and it does not break | 0.01 |
| FT | The teapot is not dropped and it breaks | 0 |
| FT | The teapot is not dropped and it does not break | 0.85 |

The conditional probability, that the teapot will break given that it is dropped, $P(b|d)$, is the result of dividing the probability of TT by the probability of TT plus the probability of TF. That makes $P(b|d)$ about 0.93. Now suppose 2.6 is a Stalnaker conditional, and we are going to evaluate it in the three branching possibilities that lie before us as we take our first step across the floor. It is trivial that 2.6 is true in TT and false in TF. But what of the FF possibility, the one we hope and intend to realize? Whether 2.6, as a Stalnaker conditional, is true or false in FF depends on whether TT or alternatively TF is the closer possibility to FF. The psychological literature on people's closeness judgements suggests that we would make this decision based on our past experience of dropping bone china on the floor or on our causal model. In these cases, and in other natural ways of judging closeness, we would decide that the TT possibility is closer than the TF possibility to FF. This judgement is clearly reflected in our behaviour: we are careful to hold the teapot securely because we judge TT is closer than TF. That means that we would judge 2.6, as a Stalnaker conditional, to be true in FF. With 2.6 true in TT and FF, the probability of 2.6, $P(\text{if d then b})$, will be the probability of TT plus the probability of FF, i.e. $0.14 + 0.85$, which is 0.99. The probability of 2.6, as a Stalnaker conditional, is consequently not the same as the conditional probability, at about 0.93.

What are the psychological consequences of the fact that the probability of the Stalnaker conditional cannot, in general, be identical with the conditional probability? We have already referred, in Chapter 1, to the growing experimental evidence that people commonly (though not invariably) give the conditional probability as the probability of the ordinary indicative conditional (see Chapter 8). This evidence would seem to count against T2 and the Stalnaker conditional. However, matters are not as simple as that. As we have just illustrated, an ordinary conditional might be a Stalnaker conditional, and people might often judge its probability to be *close to* the conditional probability, making it hard to find a significant difference between the two probabilities in an experiment. We can easily construct *artificial* examples in which the probability of a Stalnaker conditional and the conditional probability are far apart. For our teapot example, we could stipulate, in Stalnaker's formal semantics, that TF is the 'closer' possibility to FF and leave the probabilities of the four possibilities unchanged. That would make 2.6 as a Stalnaker conditional false in FF. The conditional probability would be unchanged at about 0.93, but the probability of 2.6 as a Stalnaker conditional would become 0.14, owing to it now being true in TT alone. In this artificial example, we have stipulated, for no intuitive reason, that TF is closer than TT to FF, while keeping the probability of TT much higher than that of TF. However, ordinary people's closeness judgements do not seem so detached from their probability judgements.

The experimental problem here follows from the way people appear to make both probability and closeness judgements. The two judgements are often tied together. Suppose that we want to run experiments on whether 'causal' conditionals are Stalnaker conditionals. People may judge a possibility to be very 'close' if it conforms to their causal beliefs, and for this reason, they may judge it to have a relatively high probability. They may also judge another possibility to be very 'distant' if it clashes with their causal beliefs and consequently judge it to have relatively low probability. These judgements could make the probability of a Stalnaker conditional almost equal to the conditional probability. We will return in Chapters 7 and 9 to say more about this problem, and the difficulty it makes for getting experimental evidence to separate T2 and T3 accounts of the ordinary conditional.

There is a natural deduction system for the Stalnaker conditional (Thomason, 1970) that arguably shows it to be at least close to the intuitive logic of the ordinary conditional. We also find this system attractive because it fits well with our aim of developing a theory of hypothetical thinking. In this system, there is distinction between making an assumption about the actual state of affairs and an assumption about, as we would put it, a hypothetical state of affairs. Some form of if-introduction is intuitively part of everyday hypothetical thinking, but we showed above that truth functional if-introduction is not valid for the ordinary conditional. One important difference is that only an assumption or supposition about a hypothetical state of affairs can be used in the if-introduction rule for the Stalnaker conditional.

When we make an assumption about a hypothetical state of affairs for a logical derivation, we cannot add to it premises that apply only to the actual state of affairs. (See Thomason, 1970, for the strict rules about this.) This aspect of the natural deduction system for the Stalnaker conditional enables us, for example, to avoid the second basic 'paradox' P2 of the truth functional conditional: inferring 'if p then q' from q alone. We might assume that plants will die in an actual drought, but we cannot add this to a supposition that it will rain in some alternative hypothetical state of affairs. We can quite consistently and reasonably believe in a drought that it will not rain and hence the plants will die, but by a piece of hypothetical thinking also conclude that, if it does rain, the plants will live. The paradoxical inference that 'the plants will die, therefore if it rains then the plants will die' is fortunately not permitted by this version of if-introduction.

The logic of Stalnaker conditional differs in other ways from that of the material conditional and its 'paradoxes'. Suppose 'if p then q' is a truth functional conditional. Then one of the valid rules for it is MT, i.e. inferring not-p from the premises 'if p then q' and not-q. Another valid rule for it is contraposition, inferring premises 'if not-q then not-p' from 'if p then q'. The validity of both of these rules, for the truth functional conditional, is easy to see by using truth tables or the natural deduction system for extensional logic. However, although MT is valid for the Stalnaker conditional, contraposition is invalid for it, as is easy to demonstrate by using either Stalnaker's semantic analysis or the natural deduction system of Thomason (1970).

# The T3 family

The Stalnaker conditional has many attractive features, but its probability cannot be equated with the conditional probability. In the T3 family, we can hold that the probability of the ordinary indicative conditional is identical with the conditional probability, but this requires a radical change in how the conditional is conceived (Adams, 1975, 1998; Edgington, 1995, 2001, 2003; Bennett, 2003). In T3, the meaning of a conditional is not fully given by truth conditions, but at least partly by psychological processes, those that implement the Ramsey test.

For all their differences, T1 and T2 agree that the meaning of a conditional is fully given by truth conditions, defined by reference to possible states of affairs. Recall how the meanings of conjunction, 'p & q', and of disjunction, 'p or q', are given in this way. The truth conditions of 'p & q' are given by the fact that it is true in TT and false in TF, FT, and FF. The truth conditions of 'p or q' are given by the fact that it is true in TT, TF, and FT, and false in FF. The probability of 'p & q', or of 'p or q', is then given by the sum of the probabilities of the possible states in which it is true. For T1 and T2, we can make similar statements about the ordinary indicative conditional. In T1, this conditional is truth functional. Its truth conditions are given by the fact that it is true in TT, FT, and FF, and false in TF. Its probability is the sum of the probabilities of TT, FT, and FF. In T2, the truth conditions of the ordinary conditional are given by the possible states plus a relation of relative closeness between them. Hence, in T2, as in T1, the probability of the conditional equals the sum of the probabilities of the possible states of affairs in which it is true.

Going on to T3, we would say that the conditional is true in TT and false in TF. (See Edgington, 1995, 2001, 2003, but also Bennett, 2003.) So far this does not look like a radical change from T1 and T2. However, in T3, the conditional is not given an 'outward' truth condition, or truth value, at all in FT and FF. There is a *truth value gap* in the possible states FT and FF, which are, in a sense, ignored by T3 for the evaluation of a conditional, so that T3 semantics is like the Ramsey test itself. In fact, according to T3, the psychological processes that implement the Ramsey test help to give an 'inward' meaning to the ordinary conditional. As T3 owes so much to Adams (1975, 1998), we will term the conditional in T3 the 'Adams conditional', using this phrase in a wide sense. This conditional cannot be taken as a conditional *proposition* that is always either true or false in a state of affairs. Using an Adams conditional expresses a subjective degree of conditional confidence, which equals the conditional probability. In a radical change, the probability of an Adams conditional does not equal the sum of the probabilities of the possible states in which it is true, as it is only true in TT.

As we have pointed out above, philosophers have said little about how the Ramsey test should be implemented, and this creates a hole at the heart of the T3 family. Another serious problem faced by the T3 family is specifying what negations, conjunctions, disjunctions, and embeddings of Adams conditionals mean (Adams, 1998; Bennett, 2003). After all, it does not make sense to say or write such things as 'p & P(q|p)' or 'P(P(q|r)|p)'. However, T3 is a developing family of theories, and these problems may eventually be overcome. (For important points about how T3 might be developed, see also Van Fraasen, 1976; McGee, 1989; Stalnaker and Jeffrey, 1994; Bradley, 2000.)

The only psychological work we know that pays serious attention to T3 is Johnson-Laird and Byrne (1991, p. 65). They immediately dismiss T3 because, they claim, it cannot define a notion of logically valid inference. This claim is an unfortunate mistake (Over, 1993). Johnson-Laird and Byrne were presupposing that validity can only be defined in terms of truth conditions, an inference being valid if and only if its conclusion must be true given the truth of its premises. However, Adams (1975, 1998) showed how to define validity for an Adams conditional using coherent probability judgements, or equivalently coherent uncertainty judgements. Let the uncertainty of a proposition, or alternatively of an Adams conditional, be 1 minus its probability. Then an inference in which an Adams conditional may occur is valid if and only if the uncertainty of its conclusion cannot exceed the sum of the uncertainties of its premises.

The definition of validity for the Adams conditional leads to an important result on its relation to the Stalnaker conditional (Gibbard, 1981). Consider only conditionals that do not contain conditionals as components, i.e. set aside conditionals like 'if p then if r then q'. Then an inference with an Adams conditional is valid, by Adams's definition, if and only if that inference with a Stalnaker conditional is valid, in Stalnaker's system. For example, the Adams conditional and the Stalnaker conditional are alike in that MT is valid for them but contraposition is not. This result increases the problem for psychology. We gave reasons above why people may often judge the probability of the Stalnaker conditional to be close to the conditional probability, which is the probability of the Adams conditional. Now we find that we cannot separate the Stalnaker conditional from the Adams conditional by testing people's endorsements of conditional inferences. We will, however, consider a range of psychological evidence about how people understand and use conditionals in this book, some of which is relevant to the question of whether a T2 or T3 explanation of the ordinary conditional is to be preferred.

## Conclusions

In the next chapter, we will begin to cover some of the psychological evidence against T1, the claim that the ordinary indicative conditional is truth functional. We argue, throughout this book, that the philosophical and logical objections to T1, and the psychological evidence against it, are overwhelming. However, it is another matter to decide whether T2 or T3 gives a better normative theory of indicative conditionals in natural language, or a better descriptive account of people's reasoning with this conditional and their judgements about it. We will return to this topic primarily in Chapter 9.

# 3 Indicative conditionals: the psychological evidence

In Chapter 2, we examined the writings of philosophers and logicians about indicative conditional statements. We talked a lot about the material conditional, which renders 'if p then q' equivalent to 'not-p or q'. The material conditional is truth functional. It is false when p is true and q is false, and is otherwise true. We also showed why many philosophical logicians reject the T1 hypothesis that the ordinary indicative conditional of everyday language is material. This hypothesis generates unacceptable paradoxes in natural language. It emerged that the task of replacing the material conditional with an alternative logical representation more closely mimicking the use of the ordinary conditional has been fraught with difficulty. The attempt to do so has also raised questions and issues of a distinctly psychological nature. In fact, many of the proposals that philosophical logicians have made about the ordinary conditional partly constitute, in effect, psychological hypotheses.

In this chapter, we examine for the first time the results of psychological experimentation but do not assume in our presentation that all readers of this book will have been trained in psychological research methods. Let us start with T1, a hypothesis that should be fairly easily decidable by psychological experimentation. T1 is the hypothesis that the ordinary indicative conditional is the material conditional. To put it another way, if T1 is correct, then people will form a representation of the sentence 'if p then q' that embodies the logical properties of the truth functional conditional. This being the case, we could predict how people should perform on a range of logical tasks with conditional statements. Of course, if the arguments of philosophical logicians are right, then these predictions should fail, showing that the ordinary conditional is not truth functional. Note that this approach replaces the philosophical method of a priori argumentation with the scientific method of empirical test. Note also that should T1 fail, then the psychological experiments might provide us with evidence of what alternative form of representation underlies the use of the ordinary conditional.

The research we consider falls within the *deduction paradigm*, which has been a major tool for investigating thinking and reasoning over the past 40 years or so (Evans, 2002a). The paradigm was developed during the 1960s at a time when 'logicist' thinking was strong in psychology mostly due to dominant theorizing of Jean Piaget. Psychologists assumed that logic provided the correct normative account of rational reasoning at this time and many also believed that people had a built-in logic in the mind. Hence, it made sense to investigate the extent to which people were competent in the solution of deductive reasoning tasks and to examine the extent of correspondence between their response and those that a logical analysis would dictate. For their normative frameworks and the development of suitable reasoning tasks, psychologists drew heavily upon two forms of logic: standard propositional logic and Aristotelian syllogistic logic. The former led to major study of conditional reasoning and to some study of disjunctive reasoning.

In examining psychological evidence, our difficulty will not be lack of evidence: indeed there is a wealth of studies. The problem is that psychological experiments are a lot more complex to interpret than might at first be thought and introduce numerous unexpected and often perplexing findings. When psychologists started studying deductive reasoning the approach was apparently straightforward. Give people problems that have a clear logical structure together with appropriate instructions (assume the premises, draw necessary conclusions). Then compare their answers with the normative ones. If they correspond correctly, then they are deductively competent, otherwise they are incompetent. As Evans (1982) observed in an early review of the field, however, the reality is much more complicated. The main difficulty is that no psychological task has been devised that *purely* measures logical understanding or reasoning ability. Every task has its own characteristics and features that seem to affect performance. For example, people show an apparent grasp of a logical principle on one task they are given, but seem not to understand it on another. It may take a good deal of skilled psychological experimentation to understand what lies beneath this.

In this chapter, we will keep things as simple as possible. We will consider only studies using abstract or arbitrary problem materials. This avoids, for now, the complications introduced by use of semantically rich problem content or context that very markedly affects reasoning performance. Studies of this kind will be considered in detail in later chapters. We will divide psychological studies of such abstract, indicative conditionals into two broad kinds. First we look at studies that examine by several different methods people's understanding of the truth conditions of conditional statements; that is when conditionals will be true or false. Subsequently, we will look at studies that have examined the basic inferences that people are willing to endorse or draw from conditional statements. The psychological work includes studies of conditionals phrased both as 'if p then q' and as 'p only if q', and also conditionals with negated components. Finally, we look at some studies of the development of conditional reasoning in children.

## The truth conditions of conditionals

According to T1, people should represent the content of an ordinary conditional in a way consistent with the truth table of the material conditional (see Chapter 2 and Table 2.1). Suppose we have a set of cards each of which have a capital letter written on the left and a single digit number written on the right. Then we could have an indicative conditional statement (sometimes referred to by psychologists as a 'rule' or a 'claim') that might be true or false, such as:

3.1 If there is a B written on the left then there is a 5 written on the right

Suppose we asked someone to sort through a pack of such cards, deciding whether each card would make the statement true or false. It seems clear that people might classify cards with B and 5 as true and cards with B plus some number other than 5 as false. The tricky question is how to classify cards that do not have a B on the left, but have something else, like an E. If the conditional is truth functional, then these should be classified as true cases also, regardless of whether there is a 5 on right. An alternative hypothesis was proposed by the psychologist Peter Wason, who described it as a *defective* truth table. He suggested that people would see the false antecedent cases, not as making the conditional true, but as *irrelevant* to it (Wason, 1966). The defective conditional hypothesis is of great relevance to some of the philosophical debate about indicative conditionals considered in Chapter 2. In particular, it appears to accord with the predictions for a T3 account of conditionals in which people are supposed to have 'truth value gaps' for the false antecedent cases.

Some psychologists have suggested also that people often represent 'if p then q' not as a conditional but as a biconditional, meaning 'If p then q and if q then p'. For example, they might think that the above statement actually means:

3.2   If there is a B written on the left then there is a 5 written on the right and if there is a 5 written on the right then there is a B written on the left

It can be argued that the conditional sentence form is ambiguous for abstract statements presented out of context (see Evans *et al.*, 1993, Chapter 2). This is because in everyday language we may intend either the conditional or the biconditional meaning according to context. Note, however, that biconditionals can be 'defective' in the Wason sense as well. Consider the above biconditional. The first conditional, if defective, would lead us to make the following classifications:

B & 5 (true); B & not-5 (false); not-B & 5 (irrelevant); not-B & not-5 (irrelevant)

These are the four cases usually described as TT, TF, FT, and FF after the truth and falsity of the antecedent and consequent (see Chapter 2). The second, converse conditional (If 5 then B) similarly leads to the following classifications:

5 & B (true); 5 and not-B (false); not-5 & B (irrelevant); not-5 and not-B (irrelevant)

which are the cases TT, FT, TF and FF in terms of the original conditional. So the only tricky cases are TF and FT, each of which is irrelevant by one conditional but false by the other. As the conditionals are conjoined, however, these cases must be false, leaving just FF as irrelevant in the defective biconditional. The proposed truth tables for the four hypotheses are shown in Table 3.1.

In looking at the psychological evidence we will first consider only cases when the conditional statements are affirmative. The complications introduced by negated conditionals are considered subsequently.

**Table 3.1.** Possible truth tables for conditional statements

|  | TT | TF | FT | FF |
|---|---|---|---|---|
| Representation |  |  |  |  |
| Material conditional | T | F | T | T |
| Defective conditional | T | F | I | I |
| Material biconditional | T | F | F | T |
| Defective bicondtiional | T | F | F | I |

T = True; F = False; I = Irrelevant (or indeterminate).

## Affirmative indicative conditionals

The first 'truth table experiment' performed by psychologists on conditionals seems to be that of Johnson-Laird and Tagart (1969). They asked participants to sort a set of cards similar to those we have been discussing into three piles for true, false, and irrelevant. Of the 24 participants, 19 (79%) consistently sorted according to the defective conditional pattern, just as predicted by Wason (1966). There have been a number of replications using this methodology (see Evans *et al.*, 1993), although there is the possibility that giving people 'irrelevant' as an explicit sort category might prompt them to use it. An alternative method (Evans, 1972a; Oaksford and Stenning, 1992) which avoids this problem asks people to construct examples of true and false cases exhaustively, although they are allowed to generalize ('A B with any number other than a 5 would break the rule' etc.). In this method, cases that people do not offer either to verify or falsify the statements are *inferred* to be irrelevant, without the concept of irrelevance ever being mentioned in the instructions. Results using this method also generally comply with the defective conditional hypothesis. People construct the TT case when asked to confirm the statement, TF when asked to falsify it. FT and FF cases are quite rarely offered, although the defective biconditional pattern is occasionally observed (that is, FT is sometimes given as a falsifying case).

The defective conditional hypothesis is strongly supported by these kinds of truth table experiments, but a somewhat different view is given by a task in which people are asked to list *possibilities* as in several papers by Barrouillet and colleagues (for example, Barrouillet and Lecas, 1998, 1999). Asking people to list what is impossible produces similar results to the false instruction on the truth table task. However, when people are asked to list what is *possible* given the conditional statement, they often include the FT and FF cases. This occurs quite commonly with older children and adults, even though such cases are rarely included on the conventional truth table task when people are asked to identify cases that are *true*. The distinction is significant for the mental model theory of conditionals (Johnson-Laird and Byrne, 2002) that predicts precisely this finding in its extensional account of the meaning of conditionals (see Chapter 4).

We are a little suspicious of this 'possibilities' form of the truth table task that is exclusively used by followers of the mental model theory and seems to us to obscure a distinction that the standard task reveals. If you have a defective truth table and are asked to classify impossible cases, this has to mean TF. However, possible/impossible is a dichotomy so that when classifying both true and irrelevant cases (TT, FT, and FF) with this instruction you are unable to signal their difference. 'Possible' here simply means 'not impossible'. 'True' on the other hand, cannot be assumed to mean 'not false' when 'irrelevant' is a possible third way to classify cases. What 'irrelevant' means is not entirely clear either, but we will return to that issue a little later.

The truth table task has been applied to conditionals of the form 'p only if q' as well. In standard truth functional logic, this has the same truth table as 'if p then q', making the two forms of expression logically equivalent in this logic. Of course, we do not believe that these forms are truth functional in natural language, but however that may be, they are clearly not linguistically equivalent, as becomes apparent when temporal-causal relations are introduced. For example, the statement, 'If you turn the key, the engine will start' *cannot* be expressed as 'You turn the key only if the engine will start'. Of course, you could state that, 'The engine will start only if the key is turned' but now antecedent has become consequent and vice versa, and the logical implications are different. In temporal-causal statements, 'if' must always be linked in natural discourse to the event that occurs earlier in time even when accompanied by 'only'. Hence, we could state 'People attend the theatre only if they have bought a ticket', even though it is the presence at the theatre that implies the ticket buying and not the other way around.

Truth table responses have been compared between 'if then' (IT) and 'only if' (OI) conditionals in a couple of studies (Evans, 1975; Evans and Newstead, 1977). Evans and Newstead (see also Evans and Beck, 1981) used response timing to test the temporal directionality hypothesis. In context where 'first' and 'second' had a clear temporal reference, they compared judgements for the following four kinds of statements:

Forward IT: If the first letter is T then the second letter is B
Backward IT: If the second letter is G then the first letter is Q
Forward OI: The first letter is R only if the second letter is A
Backward OI: The second letter is M only if the first letter is E

The methodology used allowed measurement of the time taken to comprehend each statement of this type and produced a very clear finding. While 'if then' conditionals were more quickly processed in the forward form, the reverse was true for 'only if' conditionals. Thus even in this arbitrary context reversing the normal temporal application of each statement increased their difficulty of understanding. However, examination of the truth table patterns showed remarkably little difference between the two forms of conditionals. The defective conditional pattern dominated for both forms with some evidence also for the defective biconditional pattern (see Table 3.1). Evans and Newstead (1977) specifically examined the data for evidence of conversion of the 'only if' from. For example, it is often suggested that people might mistake 'p only if q' for 'if q then p', which carries the converse logical implication. In fact, Evans and Newstead found little evidence to support this.

Psychologists often talk about 'material implication', meaning the material conditional, but we generally avoid this term for reasons explained in Chapter 2. The relationship represented by the material conditional can be expressed much less controversially in forms of words other than 'if then':

*Disjunction*: Either not-p or q
*Negative conjunction*: Not (p and not-q)

A recent study by Evans *et al.* (1999b) compared conditional ('if then') statements with universal statements ('Every p is a q') as well as disjunctive and negative conjunctive forms using a form of the truth table task. In one experiment, participants were asked to construct one verifying and one falsifying case. In a second experiment they were shown the four cases and asked to tick one that verified the conditional and one that falsified it. This method cannot directly show the defective truth table as people are not asked to identify *all* true and false cases. However, the first choices made are informative. The first finding was that responses were near identical when comparing 'if p then q' with 'Every p is a q'. This accords with our observation (Chapter 1) that there is nothing special about the word 'if', and that conditional thought can be expressed in other ways.

People had some difficulty understanding the disjunctive and negative conjunctive forms shown above. The statement 'either not-p or q' was confirmed fairly consistently, with 87% (81%, Experiment 2) giving the p and q case, but was falsified erratically, with only 55% (33%) identifying p and not-q. The form 'There is not both p and not-q' was falsified successfully by only 55% (56%) who found the p and not-q combination. Verifications were fairly evenly spread over the other three cases. These results illustrate a point made in Chapter 1, about how happily 'if' combines with 'not'. Whereas people have little problem understanding negations in conditionals, the introduction of negatives into other forms causes much confusion. This finding is of practical importance. Conditionals containing 'not' are often clearer ways to communicate information to people than using other connectives with 'not'. Theoretically, it is hard to explain how this can be so if T1 holds.

Summarizing to date, psychological investigations of the truth conditions of affirmative indicative conditional statements have found evidence against T1 and in favour of the defective conditional hypothesis, with some weaker support for the defective biconditional reading. A similar point can be made about 'only if' conditionals, and their difference in use when temporal relations are conveyed has also been demonstrated.

Before leaving this section, we will briefly describe the study of Evans *et al.* (1996b), who used a rather different approach to investigating the issue. In their experiments, participants were asked to construct *distributions* of the four truth table cases to represent both true and false conditionals. In one example of their tasks, participants were shown a 6 by 6 grid of 36 squares (sometimes on a computer screen, in one case physically represented). Above the grid was a conditional statement such as:

If the symbols are circles then they are yellow

The participants were then given a set of coloured shapes of nine kinds (yellow, red, or blue in colour and circular, square, or triangular in shape). The instruction was to fill in the grid using the available shapes so that the statement above was either true or false with respect to the appearance of the grid. The verification and falsification tasks were performed separately. This task is subtly but importantly different from the truth table task. From the viewpoint of T1 all that is necessary to comply with the instruction is to include at least one TF case (non-yellow circle) when representing a false conditional, and to avoid any such cases when representing a true conditional. In fact, the task is far richer than this, allowing us to simulate the kind of world that people expect to see when conditionals are true or false. For example, what balance would people expect between cases where the antecedent holds and where it does not? The experiments were conducted with 'only if' as well as 'if then' conditionals, and some illustrative data are shown in Figure 3.1.

These choices are very interesting. First, consider responses to the verification task—representing a true conditional. T1 fails because people include a small number of TF cases when modelling true conditionals. This might suggest that people do not expect a 'true' conditional to apply universally, and fits with other research showing that people interpret 'all' fuzzily or vaguely to mean 'nearly all' (Newstead and Griggs, 1984). Another possibility is that people think of *instances* of the general conditional, of the form 'If the next symbol I see is a circle then it is yellow.' These instances would be useful conditionals to assert on particular occasions if the conditional probability is high, as of course it is when there are only a small number of TF cases. This study

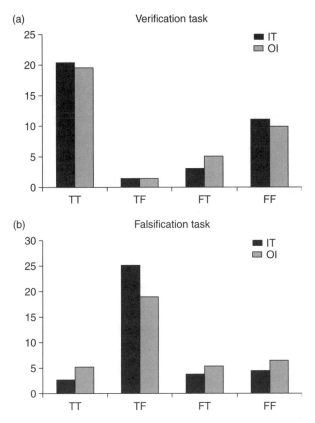

**Figure 3.1.** Data from Evans *et al.* (1996, Experiment 2, affirmative conditionals).

provides one source of evidence for a probabilistic representation of conditional statements (for more direct study of this aspect see Chapter 8) and support for a T3 theory of conditionals. However, note that people include quite a few false antecedent cases. This fits with the findings of Barrouillet and colleagues (for example, Barrouillet and Lecas, 1989, 1990) that people will list such cases as *possible*, given the conditional statement, even though they are judged irrelevant on the standard truth table task. Our interpretation of this finding is that people do not expect a conditional always to apply. In fact, the conditional would be anomalous in a situation where the antecedent was always true. We use 'if' to indicate that the antecedent *may* apply and to assert the consequent in this context. For example, if people gave only TT cases, this would represent the conjunctive 'p and q' rather than the conditional.

The representation of false conditionals is even more interesting (Figure 3.1b). Only one TF case is necessary, but most people include numerous such cases as well as a few of all of the others. This means that not only is the conditional probability, P(q|p), is high for the 'true' conditional, but the same probability is set to be *low* for the 'false' conditional, again suggesting a probabilistic representation of the conditional (see Chapter 8). This fact might also suggest a version of T3, in which to say that an instance of a general conditional is 'false' is to express low confidence in the consequent given the antecedent. It is also interesting to note how little difference there is in people's treatment of 'if then' and 'only if' conditionals. We can see immediately that evidence for the conversion hypothesis ('p only if q' read as 'if q then p') is weak. If people were converting 'only if' statements, then FT, not TF, would be the predominant falsifying case chosen. There is a small shift only in this direction, consistent with observation of Evans and Newstead (1977) that a small minority of participants may make this conversion.

Let us now return to the evidence of the defective truth table. Specifically, why do people say that false antecedent cases are irrelevant to the truth of the conditional when we give them the opportunity? This is clearly incompatible with the material conditional hypothesis (T1) and we have already noted that it fits well with the claim of T3 theorists that conditionals only have truth value when the antecedent is true. Does it, however, rule out T2 conditionals, such as that of Stalnaker? According to the Stalnaker's extension of the Ramsey test, discussed in Chapter 2, the degree of confidence we should have in a conditional can be determined when we do not believe the antecedent. The truth or falsity of the Stalnaker conditional can also be determined in this case, by considering the closest possible state of affairs in which the antecedent is true. That mechanism is such that our participants might still describe false antecedent cases as 'irrelevant', which need not mean 'indeterminate'. We illustrate the problem with an example. Consider the statement:

If you are in London then you have a good choice of theatres

According to Ramsey test, people can evaluate this conditional regardless of where we happen to be when it is asserted. No matter if they are in Durham or Plymouth (which have a poor choice of theatres) or in Paris (which has a great choice). What they do is to suppose that they are in London, making the least possible change to their background beliefs (not, for example, assuming that many London theatres have gone out of business). It could also be said that, in a sense, they 'ignore' the false antecedent possibility when they focus on the closest possibility in which the antecedent is true.

Another interesting possibility to consider is that abstract indicative conditionals of the kind discussed in this chapter are more likely to be treated in the T3 manner, due to their arbitrary nature. The example we have just used, about theatres in London, has a specific and realistic content. Followers of T2 could claim that their account applies to such realistic cases and not to abstract conditionals, such as, 'If a card has a B on one side, then it has a 3 on the other'. Suppose we are holding a card with some letter other than a B on it, say a D. What is the closest possible state of affairs in which the card does have a B on it? There is apparently no natural notion of 'closeness' that can help us to answer this question. Even if T3 does account best for responses to conditionals such as 3.1, as it appears, T2 might be better for the realistic conditionals we use every day. As we shall see later (Chapter 6) truth table experiments with realistic conditionals also quite often produce patterns *other* than that for the defective conditional. We will return to the issue of the defective truth table also in our final discussion (Chapter 9).

## Negated conditionals

As illustrated in Chapter 1, conditionals naturally combine with negations in either antecedent or consequent in everyday discourse. It is therefore reasonable to ask, what are the truth conditions of these negated conditionals? Do people treat these also as defective conditionals? Many years ago, the first author of this book (J.E.) established a methodology that has become known as the 'negations paradigm'. This involves running equivalent logical tests on four conditional statements of the form:

If p then q
If p then not q
If not p then q
If not p then not q

In fact, a number of the studies cited above used the negations paradigm, but so far we have discussed only their findings with affirmative conditionals. Introducing negations complicates the story quite considerably. The first paper of this kind was reported by Evans (1972a) also introducing the version of the truth table task where participants have to construct cases to make a statement true or false. They were given an array of cards showing coloured shapes and statements such as the following:

If there is (not) a red diamond on the left, then there is (not) a yellow square on the right

Each participant actually did four tasks, with each of the four statement types, using different lexical content. J.E. (a PhD student at the time) ran the experiment one-to-one and noticed some very striking behaviour, not at all predicted, but subsequently confirmed by statistical analysis. He noticed that people frequently picked out cards named in the conditional, such as the red diamond and the yellow square in the above example. For example, when falsifying statements, people generally constructed the TF case, but not when the conditional had the following (If not p then q) form:

If there is not a red diamond on the left, then there is a yellow square on the right

To make TF here, you would have to select a card that was not a red diamond (say a blue square) and place it to the left of one that was not a yellow square (say a green triangle). Few participants did this, however. Most simply placed a red diamond next to a yellow square—the FT case. We discussed earlier the biconditional reading of a conditional in which people might identify FT as well as TF as falsifying. But what would it mean to choose, as happens here, FT *instead* of TF as the falsifying case? That would suggest, apparently, that 'If not p then q' is read as its converse, 'If q then not p'.

In fact, this is not the correct explanation. What we are observing here is a cognitive bias affecting the responses made to the truth table task. Evans (1972a) termed this 'matching bias'. It consists of a tendency to see cases whose lexical content matches the explicit values in the conditional statement as relevant and conversely to fail to see the relevance of mismatching cases. Hence, in the above example the red diamond and yellow square appear to be much more relevant to the claim than do shapes that would negate these cases. The same effect has been shown subsequently (first by Evans and Lynch, 1973) to affect the Wason selection that we discuss in Chapter 5. The term 'matching bias' stuck, and the phenomenon was subjected to further investigation over the next 25 years prompting a major review by Evans (1998b).

This is our first use of the term 'bias' in this book, so we take a moment to clarify what this means. In the psychology of deduction it is conventional to describe as a bias any systematic *non-logical* influence on behaviour. Bias does not necessarily imply

**Table 3.2.** Matching cases and truth table cases, illustrating the negations paradigm

|                   | TT   | TF   | FT   | FF   |
| ----------------- | ---- | ---- | ---- | ---- |
| If B then 3       | B3   | B7   | N3   | G4   |
| If p then q       | pq   | p¬q  | ¬pq  | ¬p¬q |
| If G then not 6   | G9   | G6   | T4   | A6   |
| If p then not q   | p¬q  | pq   | ¬p¬q | ¬pq  |
| If not R then 5   | M5   | B6   | R5   | R9   |
| If not p then q   | ¬pq  | ¬p¬q | pq   | p¬q  |
| If not E then not 1 | K7 | D1   | E3   | E1   |
| If not p then not q | ¬p¬q | ¬pq | p¬q  | pq   |

irrationality, because logic does not necessarily define rationality (see Evans and Over, 1996a). The non-logical nature of matching bias is, however, clearly demonstrated by use of the negations paradigm.

To understand the nature of the negations paradigm, it is worth giving Table 3.2 a little study. This shows truth table cases for the four statement types both symbolically in terms of p and q (where ¬p means 'not p' etc) and with examples using letter–number pairs. A double matching case—pq—is one where both the letter and number in the statement can be matched to produce the required truth table case. A double mismatching case, ¬p¬q, is one where both letter and number have to be altered to something else. The logical cases are TT, TF, FT, and FF and matching cases are pq, p¬q, ¬pq, and ¬p¬q. If we use only affirmative conditionals, the two are *confounded*, for example p¬q always represents the TF case. By using all four conditionals with negated components, however, the two ways of defining cases are counterbalanced with respect to each other, or orthogonal. Suppose we averaged responding over a given logical case—say FT. We see that over the four conditionals, this corresponds to each of the matching cases, pq, p¬q, ¬pq, ¬p¬q, on one conditional or another. The same is true for the other truth table cases. Conversely, suppose we average choices of a particular matching case, say p¬q. We find that this corresponds on some conditional or other to each of the four logical cases, TT, TF, FT, and FF.

The presence of matching bias creates a complication for the interpretation of research on affirmative conditionals, discussed earlier. Matching and logical case are confounded, so that we cannot be sure which is determining choices. One way to deal with this, is to average responses to the four logical cases across the four statement types, so that matching bias is balanced overall. One of the larger data sets available where participants were asked to rate cases as true, false, or irrelevant is that of Evans and Newstead (1977). Table 3.3 shows percentage ratings of each case, averaged across the four statement types so that matching bias is controlled. The pattern that emerges still broadly supports the defective conditional hypothesis. The dominant choice for TT is true and for TF is false for both 'if then' and 'only if' conditionals. Although irrelevant responses are far more common on the FT and FF cases, something more is going on here. FT cases are quite often classified as false (especially for 'only if') giving some

**Table 3.3.** Truth table judgements (%) from the study of
Evans and Newstead (1977), averaged across matching
cases

|  | TT | TF | FT | FF |
|---|---|---|---|---|
| If then |  |  |  |  |
| True | 97 | 7 | 6 | 28 |
| False | 2 | 84 | 48 | 23 |
| Irrelevant | 1 | 9 | 46 | 49 |
| Only if |  |  |  |  |
| True | 87 | 6 | 8 | 32 |
| False | 4 | 75 | 67 | 23 |
| Irrelevant | 9 | 19 | 25 | 45 |

evidence of biconditionality. FF cases do seem irrelevant and, although they are sometimes classed as true or false, there is no consistent preference.

Hence, matching bias complicates the story, but not hopelessly so. It is interesting in its own right. We observed in Chapter 1 that *not* as well as *if* is closely linked to hypothetical thought. In the heuristic-analytic theory of Evans (1989) each word is linked to a heuristic that determines perceived *relevance* of cases. Cases seem more relevant when the antecedent is true (the if-heuristic) and when they match (the not- or matching-heuristic). We briefly consider the cause of the matching bias effect. At the time of writing, there are two main rival accounts of the phenomenon. We favour the view that it is linked with the use of implicit negation (Evans, 1998b), for which strong evidence has been provided in several studies (Evans, 1983; Evans *et al.*, 1996a; Evans and Handley, 1999). Looking at Table 3.2, we can see that mismatching cases are produced by implicit negations, for example 7 stands for not-3. The studies cited above have shown that when explicit negations are used, for example in the truth table task, the matching bias effect disappears. For example, we can modify the task so that the case is described in a sentence. Say that the conditionals always refer to a letter–number pair as in the table and the conditional is:

If the letter is not R then the number is 5

The TF case could be described as:

The letter is B and the number is 6

for one group of participants given implicit negations. However, it could be described for another group using explicit negation as follows:

The letter is not R and the number is not 5

Notice that in the explicit negation group *all* cases match, that is refer to letters and numbers named in the conditional statement. The effect of this manipulation was to remove *entirely* the matching bias effect and also to increase correct true and false classifications of the TT and TF cases. In spite of this dramatic evidence, there is support for an alternative theory based on expected information gain (Oaksford and Chater, 1994).

According to this view negative assertions are difficult to process because they usually refer to large contrast classes and provide low information. For example, if I state that 'the letter is B' this is highly informative as it eliminates 25 of 26 possibilities of what the letter might be. If I say instead 'The letter is not A', I provide little information, eliminating only one of 26 possibilities. This difference in information value could lead to people paying more attention to items defined by explicit affirmative reference than to those representing reference classes of negated assertions. If this is correct, then matching bias should be reduced when contrast class sizes are small. A recent study manipulated implicit negation and contrast classes independently to try to separate these accounts (Yama, 2001), but found evidence for both explanations! For recent discussion of these and other related findings see Evans (2002b) and Oaksford (2002).

Before leaving the topic of negated conditionals, it is pertinent to ask a further question. Matching bias is strongly established for both 'if then' and 'only if' conditionals. But does it depend upon conditional statement forms or can it be shown with other propositional connectives? There was considerable doubt over this issue until the recent study of Evans *et al.* (1999b). In this study, mentioned earlier, they compared conditionals with universal, disjunctive and conjunctive forms, employing the full negations paradigm with each. For example, disjunctive were expressed as 'Either p or q', 'Either p or not q', 'Either not p or q', and 'Either not p or not q'. Although the patterns differed somewhat, there was clear evidence of matching bias with disjunctive and conjunctive forms, suggesting that conditionality is not required.

From the viewpoint of discovering how people understand the truth conditions of 'if', the matching bias effect is something of a headache. This is the kind of finding we had in mind earlier when referring the difficulties of interpreting psychological experiments. Psychology has its own version of Heisenberg's uncertainty principle, in that any particular task you choose to measure a mental process is likely to affect that process in some way. The presence of matching bias makes if very difficult, for example, for us to decide whether the truth conditions for' if p then q' are different from those for 'if not p then q', as the bias affects participants' choices in different ways on the two statements. We will encounter similar complications in studies based on inference tasks, discussed below. We ask our readers to be patient, as the true nature of 'if' will gradually emerge from the evidence we review in this book. For the moment, we can certainly say that the evidence weighs against T1, and there is overall evidence for some kind of 'defective' truth table.

## Conditional inference

The great bulk of psychological research on conditionals has used the conditional inference task. It is a mainstay of logic textbooks that the material conditional supports two valid inferences and is prone to two invalid inferences. Actually, there are infinitely many inferences that might be made with conditional statements, but the standard paradigm focuses on the following four, each of which asserts the truth or falsify of one component of the conditional and draws a conclusion about the truth or falsify of the other. These four inference forms are the following.

*Modus Ponens (MP)*
> If p then q; p, therefore q

*Example*:
> If there is an A on the left then there is a 3 on the right
> There is an A on the left
> Therefore, there is a 3 on the right

*Modus Tollens (MT)*
> If p then q; not-q, therefore not-p

*Example*:
> If there is a G on the left, then there is a 1 on the right
> There is not a 1 on the right
> Therefore, there is not a G on the left

*Affirmation of the Consequent (AC)*
> If p then q; q, therefore p

*Example*:
> If there is a T on the left, then there is a 4 on the right
> There is a 4 on the right
> Therefore, there is a T on the left

*Denial of the Antecedent (DA)*
> If p then q; not-p, therefore not-q

*Example*:
> If there is a C on the left, then there is a 7 on the right
> There is not a C on the left
> Therefore, there is not a 7 on the right

An argument is (logically) valid if and only if there is no logical possibility in which its premises are true and its conclusion false. A (logical) fallacy is an argument that is not (logically) valid. In a valid argument, the conclusion is logically necessary given the premises, and in a fallacy, the conclusion is not logically necessary given the premises. Hence, the standard instructions in psychological experiments, using the conditional inference task, ask whether conclusions necessarily follow from given premises, which are to be assumed true. MP and MT are valid under T1, T2, and T3 accounts of the conditional. Similarly, DA and AC are fallacious under all theories unless a biconditional relationship is assumed. Hence, the studies reviewed in this section should not be expected to discriminate between these accounts. The study of conditional inferences both with the abstract conditionals reviewed here, and the thematic conditionals discussed in Chapter 6, does, however, provide interesting information about conditional thought.

The validity for the material conditional of MP and MT, and the invalidity of AC and DA, follows from the truth tables in Table 2.2. We should note as well that MP and MT are valid, and AC and DA invalid, for the Stalnaker conditional, as can be seen from the semantic analysis of this conditional (Chapter 2 and Stalnaker and Thomason, 1970). We also noted earlier that some psychologists have proposed a 'chameleon' theory, according to which an ordinary conditional is often interpreted as a biconditional. If a conditional is represented as a biconditional, all four forms of inference become valid for both the material and the Stalnaker conditional.

An intuitive derivation of MT can be made by a suppositional inference in a natural deduction system. Consider the two premises of:

If there is a G on the left, then there is a 1 on the right
There is not a 1 on the right
Therefore, there is not a G on the left

We can argue as follows. Suppose that there is a G on the left. Then there must be 1 on the right (MP). But there is not a 1 on the right (premise 2). Hence, we have a contradiction. So the supposition that led to this contradiction must be false: there cannot be a G on the left. This instance of a reductio ad absurdum argument is valid both for the material conditional and the Stalnaker conditional (Thomason, 1970).

Before we look at some psychological data, it is important to note two more things. From a logical point of view, we could only expect one of two patterns of inference to emerge. People should either endorse just MP and MT (conditional) or all four inferences (biconditional) according to how they interpret the statement. Second, the inference task cannot discriminate between full and defective versions of the truth tables (Table 3.1). MP and MT follow from the fact that TF is false (all truth tables) and DA and AC follow from the fact that FT is false (in the table for both full and defective biconditionals). For example, it is not necessary to represent FF as a true possibility in order to make MT, as assumed by the mental model theory of conditional inference (see Chapter 4).

So what do participants in the psychological experiments do? Evans *et al.* (1993) collated results from several studies using affirmative indicative conditionals with abstract problem content (their table 2.4, p. 36) and found quite variable outcomes. MP endorsement rates ranged 89–100% across studies; DA 19–73%; AC 23–75%, and MT 41–81%. One finding that is consistent across studies, and a basic phenomenon for any psychological theory of conditional inference to explain, is that MP is made significantly more frequently than MT. There is some ambiguity in the literature as to whether AC also occurs more often than DA. The overall level of both DA and AC inferences varies dramatically between studies, probably because different instructions, materials, or context may tend or not tend to suggest a biconditional reading.

In view of this rather confusing picture, we look at the results of one recent study of conditional inference that used large samples and applied two different methods. Evans *et al.* (1995) used abstract letter–number conditionals in two experiments. In the first experiment, they gave the conditional premise, minor premise and the conclusion and asked participants to decide whether the conclusion necessarily followed. For example, the following DA inference:

If the letter is B then the number is 7
The letter is not B
*Conclusion*: the number is not 7          YES          NO

Participants were instructed in this experiment that their task was to decide whether or not the conclusion necessarily followed from the statements. This is known as an evaluation task. In Experiment 2, everything was similar except that a production task was used. This was presented as follows:

If the letter is B then the number is 7
The letter is not B
*Conclusion*............................

Here the instruction was: 'If you think a conclusion necessarily follows the please write it in the space provided. If you think that no conclusion follows, then please write NONE'.

Evans *et al.* (1995) looked at three forms of expression of the conditional: 'if then' (IT), 'only if' (OI) and reverse if (IF). We have met the IT and OI forms before. The IF form simply involve stating the consequent before the antecedent, as in q, if p, or 'The number is 7 if the letter is B'. All three forms are logically equivalent. Evans *et al.* also employed the negations paradigm, but we consider first only the findings with affirmative conditionals (Table 3.4). This is one of the studies that shows very high rates of the 'fallacies', DA and AC, although there is nothing in the wording of the problems that obviously suggests a biconditional interpretation. The authors also observed significantly higher rates of AC than DA inferences, although this is not consistently reported in the literature. MT rates were higher than is often observed, but still well below those for MP.

Clearly, these findings do not fit either of the patterns of logical inference that should follow from the truth tables. In other words, the reasoning is not truth functional. As pointed out earlier, people ought either to make MP and MT (conditional) or all four inferences (biconditional). Instead, they make MP more than MT and AC more than DA. These differences have to lie in the mental processes that people employ when they try to make these inferences. The MP/MT difference is a key finding that psychologists have tried to explain, as we shall see in Chapter 4.

Evans *et al.* (1995) also employed the negations paradigm throughout. Just as with the truth table task, the use of negatives in conditional inference problems demonstrates a cognitive bias in conditional inference that complicates the story. In this case the main effect is that of negative conclusion bias, or double negation first demonstrated by Evans (1972b). Consider the following problem that psychologists normally describe as an MT inference:

If the letter is not J then the number is 8
The number is not 8
Therefore, the letter is J

Logicians reading this might well say that the conclusion stated does not follow simply by the inference rule of MT—and they would be right. The conclusion that follows by MT is that the letter is 'not not a J'. In order to reach the conclusion stated, we need to apply a second rule of double negation elimination, which is that not not P implies P. This makes a difference, as inference rates are consistently lower on problems

**Table 3.4** Conditional inference rates (%) in the study of Evans *et al.* (1995), affirmative conditionals only

|  | MP | DA | AC | MT |
|---|---|---|---|---|
| Experiment 1 |  |  |  |  |
| IT | 98 | 79 | 88 | 74 |
| OI | 95 | 79 | 100 | 88 |
| IF | 90 | 70 | 91 | 72 |
| Experiment 2 |  |  |  |  |
| IT | 94 | 65 | 88 | 71 |
| OI | 92 | 76 | 94 | 88 |
| IF | 94 | 65 | 84 | 71 |

where such a double negation is required. This applies not only to MT problems but to DA inferences as well such as:

If the letter is R then the number is not 5
The letter is not R
Therefore, the number is 5

Although DA is not a valid rule of inference, people often reason as if it were and then are less likely to draw the inference when they hit the double negation problem (not not 5, therefore 5). This effect is really quite large. The reason for this 'negative conclusion bias' is that the inferences requiring double negation elimination always end up drawing an affirmative conclusion, and those not requiring it end up delivering a negation conclusion which is therefore more frequently endorsed. Figure 3.2 shows inference rates in the study of Evans *et al.* (1995) broken down by conclusion polarity for both DA and MT inferences. It will be seen that substantially more are drawn with negative conclusions. There is some evidence that a weaker negative conclusion bias also applies to AC inferences, even though an obvious double negation explanation is not available (Schroyens *et al.*, 2001). This effect is clearly one for which psychological theories of conditional inference must provide an explanation.

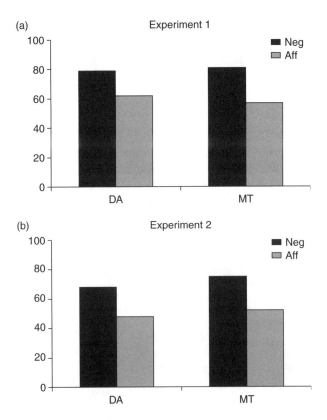

**Figure 3.2.** Percentage acceptance/production of inferences in the study of Evans *et al.* (1995) as a function of the polarity of the conclusion.

**Table 3.5.** Conditional inference rates (%) in the study of Evans *et al.* (1995), averaged across conditionals with and without negated components

|                | MP | DA | AC | MT |
|----------------|----|----|----|----|
| Experiment 1   |    |    |    |    |
| IT             | 96 | 65 | 88 | 62 |
| OI             | 93 | 72 | 98 | 76 |
| IF             | 94 | 74 | 90 | 74 |
| Experiment 2   |    |    |    |    |
| IT             | 93 | 62 | 87 | 64 |
| OI             | 87 | 64 | 89 | 70 |
| IF             | 89 | 64 | 81 | 56 |

Polarity biases in the conditional inference task introduce precisely similar complications to matching bias on the truth table task. Once again, however, we can use the negations paradigm to control for these biases. Table 3.5 shows the inference rates for the Evans *et al.* (1995) study averaged across the four forms in which negative components are rotated. This means that each inference will be associated with equal numbers of affirmative and negative conclusions and not be confounded by polarity as are the data for affirmative conditionals only (Table 3.4). Table 3.5 is directly analogous to Table 3.3 where the same technique was used to control for matching bias. Comparing Table 3.5 with Table 3.4 we see that MP and AC rates are similar but that DA and MT rates are somewhat lower. This is because conclusion polarity had little effect on the former inferences in this study, but favoured DA and MT with affirmative statements. The averaged data show a much clearer preference for MP over MT and for AC over DA.

The four inferences can be divided in terms of their directionality instead of their logical validity. Let us call MP and DA forward inferences (antecedent to consequent) and AC and MT backward inferences. The directionality hypothesis of Evans (1977, 1993) should predict more forward inferences for 'if then' and more backward inferences for 'only if'. The reverse 'if' statements, placing the consequent first, might also be expected to induce more backward inferences. Evans *et al.* (1995) found some weak support for these predictions on the overall inference rates (Table 3.5). For example, there were significantly more AC and MT inferences with the 'only if' form in Experiment 1 (see also, Evans and Beck, 1988; Evans *et al.*, 1999b, Experiment 3, for other tests of this hypothesis).

Summarizing the findings of conditional inference tasks, with abstract indicative conditionals, we can see that people's inferences do not consistently conform with any of the possible truth tables discussed earlier (Table 3.1). Note that although we recorded strong philosophical objections to the truth functional or material conditional in Chapter 2, that does not mean that the findings of these psychological experiments can be directly accounted for by any proposed alternative, such as the Stalnaker conditional. The actual patterns of inferences observed, both on the abstract tasks reviewed here and thematic conditional tasks to be discussed in Chapter 5, require the development of a specific psychological account of the pragmatic and reasoning processes involved. We note here the phenomena that need to be explained.

Although DA and AC inferences are frequently endorsed in most studies, suggesting biconditionality, a closer look at the data shows that there is much more going on than this. We must explain why people endorse more MP than MT inferences, and correspondingly more AC than DA inferences. We must also account for polarity biases. We can further require our psychological theory of indicative conditional reasoning to account for findings on the truth table task, discussed earlier, such as matching bias and tendency towards defective truth tables.

## Development of conditional reasoning

How do young children interpret conditional statements and how does this change as they grow older? Most developmental studies have used the conditional inference paradigm and we discuss these below. Very few seem to have employed the full truth table task, although one such was reported by Paris (1973) who found that young children did something rarely observed with adults. They tended to rate TT as true and all other cases as false, effectively treating 'if p then q' as though it meant 'p and q'. This conjunctive interpretation of 'if' could indicate a lack of facility for hypothetical thinking at this age. More recently, Barrouillet and colleagues report several studies in which children of different ages are asked to indicate which cases are or are not possible (Barrouillet and Lecas, 1998, 1999; Barrouillet et al., 2000). We discussed this possibilities form of the task earlier, indicating that it produces somewhat different findings from the standard truth table task.

These studies show some clear developmental patterns. For example, Barrouillet and Lecas (1998) in one experiment asked children in three groups at grades 3, 6, and 9 (aged about 8, 11, and 15 years) to say for each truth table case whether it was impossible given the conditional statement. Everyone, at all ages, classified TF as impossible and no-one classified TT as impossible. The frequency of classification of FT and FT as impossible, however, declined quite sharply with increasing age. In a second experiment, in which separate participants of similar ages were asked what was possible, the results were complementary. The interpretation offered by these authors is that there is a developmental progression from conjunctive (only TT is possible) to biconditional (TT and FF are possible) to conditional (only TF is impossible) readings. This is linked to Barrouillet's mental model account of conditional inference, discussed in Chapter 4. He has shown that it is not age as such, but increasing working memory capacity (correlated with age and general intelligence measures) that predicts the pattern of interpretation observed (Barrouillet and Lecas, 1999).

The great bulk of studies on children's conditional reasoning have used the conditional inference paradigm (see Evans et al., 1993, pp. 39–42). One typical finding is that the endorsement of conditional 'fallacies'—DA and AC—tends to decrease with age, although these can be still very high with adult groups as we saw earlier. Markovits and colleagues (for example, Markovits et al., 1998) have conducted a number of studies on this effect using contextual manipulations (see Chapter 6) to support a theoretical account in terms of development of associative and semantic memory systems. The idea is that older children and adults are more easily able to retrieve from memory counterexample cases that would block the fallacies. The reason that adults sometimes show high rates of DA and AC—as in the study of Evans et al. (1995)—could well be because

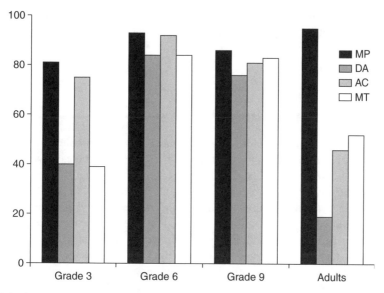

**Figure 3.3.** Conditional inference rates as a function of age, in the study of Barrouillet *et al.* (2000).

the materials are highly abstract and arbitrary, preventing access to this semantic blocking effect. Recently, Markovits and Barrouillet (2002) presented an integrated developmental account built around mental model theory.

A common puzzling finding with children's conditional reasoning is that the valid MT inference is shown to be more frequently endorsed by older children than by adults. This illustrates the difficulty of equating deductive competence with logical performance as no one can plausibly argue that children acquire a competence that they later lose. One has to look at the whole pattern of inference. The findings of Barrouillet *et al.* (2000), Experiment 3 are fairly typical in this regard (see Figure 3.3). Taken in isolation, MT rates are seen to rise to a peak in intermediate age groups (11–15 years) and then drop again in adults. Taken holistically, however, we can see that these intermediate groups make all four inferences with high frequency. Barouillet *et al.* offer the interpretation, consistent with Barouillet and Lecas (1998), discussed above, that this indicates a developmental pattern from conjunctive (MP, AC) to biconditional (MP, DA, AC, MT) to conditional (MP, MT) patterns of inference. The difficulty they have with this scheme is that the overall pattern of adult inference rates shown in Figure 3.3 does not fit well to the proposed 'conditional' representation. Not only is MP much more common than MT, but AC is similarly much more frequent than DA. An analysis of individual differences casts light on this. There is a substantial subgroup of the adults who end up being classified as *conjunctive* reasoners, apparently regressing to their early childhood. The other adults, about 50%, mostly show the conditional (MP, MT) pattern, with the adolescent biconditional pattern disappearing.

These results appear very puzzling. It seems to us inconceivable that any adults would represent a conditional statement as a conjunction, as this would effectively render them incapable of hypothetical thought. However, Evans *et al.* (2003b; discussed in detail in Chapter 8; Oberauer and Wilhelm, 2003) in a study where participants were asked to judge the probability of conditional statements also reported an individual difference analysis in which a similarly sized subgroup of adult participants appeared to equate the conditional with a conjunction.

## Conclusions

In this chapter we have reviewed the basic evidence of psychological experiments on reasoning with indicative conditionals. We have told only part of the story, however, as we have limited ourselves to studies using abstract or arbitrary problem materials. Later in the book we will examine effects of problem content and context—which are profound—and also look at studies of indicative conditionals using the Wason selection task (Chapter 6). What, however, can we conclude on the evidence so far?

To take the last evidence first, developmental studies show that understanding 'if' as a conditional connective is difficult for children and takes time to develop. Young children seem to interpret conditionals as conjunctive statements and older ones as biconditional statements (although this can be blocked with realistic conditionals statements of the right type). The biconditional pattern does not have to indicate truth functional reasoning, however. It can simply indicate a superficial reading that p and q go together. If you have one you have the other; if you do not have one, you do not have the other. As we have seen throughout this chapter, however, adult participants do not respond with consistently valid conditional reasoning. For example, they make MP much more readily than MT and AC more often then DA. If they were strictly logical, then believing that the p cannot occur without q (as the truth table tasks generally indicate) should lead to both MP and MT inferences, and believing the converse should lead to both AC and DA.

The conditional inference data are not consistent with the material conditional and neither are the attempts to measure directly the perceived truth conditions, discussed at the start of the chapter. From these studies we learned that generally people regard cases where the antecedent is false to be irrelevant to the truth of the conditional statement. However, when asked to model possible states of affairs in which conditionals hold, they include many such cases. Conditionals, in their very nature, should only apply some of the time. We also found that the presence of negated propositions, although naturally expressed in the conditional form, have a marked influence on the perceived relevance of logical cases also. Matching bias is a robust and reliable finding in this literature.

The reader might think that these rather complex and puzzling findings are due to the use of abstract or arbitrary problem materials in the experiments reviewed. These kinds of conditionals do not fall within 'natural' discourse. However, the major psychological theories of conditional reasoning, considered in the next chapter, propose a basic meaning for the connective 'if' that should be domain-independent. Both rule- and model-based approaches, as we shall see, have tried to specify a basic semantics for conditionals, and then explain content and context effects by *additional* proposals about pragmatic factors. Much of the writings of philosophical logicians on conditionals have likewise tried to establish a basic semantics for conditionals. Right or wrong, these are the standard approaches and should be able to give an account of how people assess and reason with abstract conditional statements.

# 4    The mental logic and mental models of indicative conditionals

Psychological accounts of the conditional are usually classified as being of two distinct types: mental logic theories and mental model theories. We will also briefly classify them, as far as that is possible, using the T1, T2, and T3 families, introduced in Chapter 2. Mental model theory is really a kind of mental logic, and in particular the logic of its 'basic' conditional is that of the material conditional. The mental model theory of the 'basic' conditional is the best psychological example of the T1 family. However, we begin with an exposition of what are usually called mental logic theories, based on inference rules. There are two major theories under this heading. One gives us what we call the Rips and Marcus conditional, and the other what we call the Braine and O'Brien conditional. The Rips and Marcus conditional is in the T2 family. There are too many logical problems with Braine and O'Brien's account to classify it strictly, but it has some similarities to both T2 and T3. After this, we will take a close look at the conditionals, the 'basic' and 'non-basic', of mental model theory. We call these Johnson-Laird and Byrne conditionals. The question is whether one of these psychological theories has a satisfactory account of the ordinary, natural language indicative conditional.

## The Rips and Marcus conditional

Rips and Marcus (1977) looked for a psychological 'analogue' for the 'possible worlds' in Stalnaker's formal T2 semantics for ordinary conditionals, discussed in Chapter 2. They suggested that what fulfil this role are suppositions that people naturally make. Suppose that we are feeling rather cooped up but we think to ourselves:

4.1    If we go for a walk on the moor, then we will feel much better

As we explained in Chapter 2, this conditional is true or false, in Stalnaker's semantics, if and only if its consequent is true in the possible state of the world which differs minimally from the actual state, but in which its antecedent is true. More informally, we can say that the conditional is true when we feel much better in the 'closest' or 'nearest' possible state of the world, to the actual state, in which we go for a walk on the moor. We may conclude that it is our duty to stay at work, but we could quite easily get out for the walk. In that nearby world, we will feel much better, and that makes the conditional true. Some other time, perhaps when we have flu, we do not feel better in the nearest possible state of the world in which we are anywhere else except in bed, and the conditional is false, given Stalnaker's semantics.

Rips and Marcus (1977) describe an ideal mental process of constructing 'suppositions' that they hold is an 'analogue to' or 'mimics' Stalnaker's 'possible world' semantics. It could also be said that this mental process is a kind of unbounded Ramsey test. They suggest that we can produce a ranking of the mental propositions in our 'data base' of beliefs about the current state of affairs as we take it to be. Of the highest rank are to be beliefs that we can most easily give up hypothetically. An example would be our belief that there is a meeting that we should go to. This meeting may be cancelled, and then we will be free to go up to the moor. Farther down in the ranking would be our belief that the weather is good on the moor. We quite firmly hold on to this belief because the weather forecast was for an exceptionally sunny day. There is no reason to give up this lower ranking belief merely because we are supposing that we will go for a walk on the moor. Our performing that action would not cause a change in the weather. Much farther down in the ranking is our scientific belief about what causes a sunny day to become rainy.

Rips and Marcus claim that, to evaluate our example conditional, we would begin, as in the Ramsey test, with the antecedent as a hypothetical 'seed' proposition. The goal is to construct a consistent set of propositions from this antecedent seed. We start the process by adding the seed, which in our example is of course that we go for a walk on the moor, to our beliefs of the highest rank. If this set of propositions is consistent, it becomes the 'current supposition'. If it is inconsistent, we form its largest consistent subset containing the seed, and that becomes the current supposition. We ideally proceed in this way through all the ranks until these are exhausted. At that point, we have a large consistent set: this is our 'final supposition' for evaluating the conditional. If its consequent, that we feel much better, is in that final supposition, then the conditional is true. Otherwise the conditional is false. In our example, the conditional would turn out to be true. That is because our feeling much better, rather than its negation, is consistent in this context with the hypothetical walk and the beliefs we would retain as we went through the mental process Rips and Marcus describe.

This mental process might construct more than one consistent set, so that the final supposition would be a set of consistent sets. These various sets might, for instance, specify how long a walk we took. Perhaps a very short walk would not make us feel better but longer ones would. In this case, we would judge that the conditional was probably true. In general, the probability of a conditional would depend on the ratio of consistent sets containing its consequent to the total number of consistent sets containing its antecedent. In this way, Rips and Marcus related the probability of an ordinary conditional, P(if p then q), to the conditional probability of its consequent given its antecedent, P(q|p). Unfortunately, the suggested connection was not investigated experimentally for many years. But it has become much more important as a result of recent research, which we discuss in Chapter 8.

The procedure for evaluating conditionals proposed by Rips and Marcus has some similarities to the theory of mental models (Johnson-Laird and Byrne, 1991, 2002), which we will cover in detail later in this chapter. Consider this conditional with a conjunctive antecedent:

4.2   If we go for a walk and sprain our ankle, then we will feel worse

To follow Rips and Marcus procedure, we would begin with the antecedent seed:

We go for a walk and sprain our ankle

And to construct a consistent set from this seed, we would obviously have to add both conjuncts of this conjunction to the expanding supposition set. We can think of this set as represented from left to right on one line:

We go for a walk    We sprain our ankle

As we shall see, that the above would be the mental models for the conjunctive seed. As another example, consider a conditional with a disjunctive antecedent:

4.3   If we go for a walk or have a swim then we will better

We then have the disjunctive seed:

We go for a walk or have a swim

And this disjunction divides the procedure into two consistent sets, which can be thought of as represented on separate lines:

We go for a walk

We have a swim

Another line could represent the third possibility that the disjunction holds because we both have a walk and go for a swim. Again, as we shall see below, the representation of these three possibilities would be mental models for the disjunctive seed.

Rips and Marcus were in advance of their time in relating conditionals to conditional probability. They did this by pointing out that their suppositional process will sometimes generate more than one consistent set, as in cases of conditionals with disjunctive antecedents, as we have just illustrated. However, Rips and Marcus seem to have presupposed that the generated consistent sets would be equally probable. A natural, and indeed necessary, extension of their approach is to allow for different probability weightings among the consistent sets. For example, it might be that the consistent set in which we go for a walk should be judged much more probable than the consistent set in which we have a swim. This would be so if we almost always went for a walk rather than a swim when we needed exercise. This extension would allow Rips and Marcus to account for a wider range of probability judgements about conditionals.

There is much of stimulating interest in the theory of Rips and Marcus. They are inspired by Stalnaker's formal semantics and apparently accept its natural deduction rules as the mental logic for inferring purely logical relations (Stalnaker and Thomason, 1970; Thomason, 1970). They describe a procedure for evaluating contingent conditionals and making probability judgements about them. But there are decisive objections. People cannot get entire 'possible worlds' into their heads, let alone sets of them. Even if they had only true beliefs, their mental representation of the actual world would be severely limited, and they can devote even less mental space to representing possibilities. The ideal procedure of constructing consistent sets is too close an 'analogue' of 'possible worlds'. It is impossible for finite beings, let alone ones as limited as we are, to construct such consistent sets. Above the level of the most elementary sentences, there is no effective procedure for deciding whether or not a set of sentences is consistent. And even where algorithms do exist for determining the consistency of some elementary sentences, these could take millions or billions of years to apply. These are technical, logical facts about consistency (see Oaksford and Chater, 1991, on their relevance to cognitive psychology). But it is clear, to begin with, that it is practically impossible for people to rank their beliefs in terms of their importance for describing some given situation. People cannot, in general, even call all their beliefs about some topic to mind, when asked to do so. Unfortunately, Rips and Marcus have not developed their ideas to make them more psychologically plausible, or updated them in light of the rapid increase in experimental results about conditionals in recent years.

## The Braine and O'Brien conditional

Braine and O'Brien (1991) proposed the existence of a kind of mental natural deduction system, with introduction and elimination inference rules for the ordinary conditional and other sentential connectives. Their theory was, in some respects, similar to Stalnaker's version of T2, as they acknowledged. However, they appear confused about what Stalnaker (1968) clearly and rightly distinguished, and did not fully understand Stalnaker's semantics for his conditional. Their account of the meaning of an ordinary conditional has some similarities to T3.

Braine and O'Brien (1991) tried to make the inference rules for their conditional do too much. These rules were supposed to fulfil three roles, which Stalnaker rightly distinguished. First, the inference rules were meant to specify which inferences are valid, where validity is the formal logical concept. That is, these rules were intended to tell us which inferences, such as Modus Ponens (MP) and Modus Tollens (MT), are valid in virtue of their logical form (pp. 182–183). This role should include the logically related object of inferring that certain conditionals, e.g. 'if p & q then p', are logical truths because of their logical form. Second, the inference rules were supposed by Braine and O'Brien to supply all the semantics necessary for the conditional: no reference to possible states of affairs was held to be necessary or desirable (pp. 197–200; O'Brien and Bonatti, 1999; Over and Evans, 1999). Third, the rules were to tell us when to assert a particular contingent conditional, such as the one above about going for a walk (4.1). In this role, the rules were to act, in effect, as a kind of Ramsey test. Braine and O'Brien (1991) do not refer to the Ramsey test by name, but discuss Stalnaker's version of it (p. 198).

Two rules in Braine and O'Brien were supposed to be primary in giving the ordinary indicative conditional its meaning. Their position here could be thought of having some similarity to what the T3 family holds about the Ramsey test as part as the meaning of this conditional. The primary rules for Braine and O'Brien were MP and what they called 'conditional proof' or what we called, in Chapter 2, if-introduction. Of course, these rules have to be used in general along with other inference rules, for negation, conjunction, and disjunction, which were to have an indirect role in the meaning of the Braine and O'Brien conditional (see Braine and O'Brien, 1998b for a comprehensive list of inference rules). Consider the trivial derivation in which we infer 'if p & q then p' as a logical truth. First, we assume 'p & q' and then infer p by and-elimination. The second and last step is to infer 'if p & q then p' by if-introduction, or conditional proof as Braine and O'Brien would term it. It is the total set of rules for Braine and O'Brien's mental logic that is supposed to give their conditional and all the other logical connectives distinctive meanings.

In technical terms, Braine and O'Brien (1991) tried to give an inferential-role semantics for their conditional (see also O'Brien and Bonatti, 1999). Their mental logic is not the same as classical logic because of modifications and restrictions they placed on their rules. For example, they specified that, in their logic, nothing at all validly follows from a contradiction, of the form 'p & not-p'. So when a contradiction appears in an attempt to infer a conditional by 'conditional proof', the derivation has to stop in Braine and O'Brien's system and nothing follows. The derivation also has to stop when a conclusion is inferred that is inconsistent with a preceding proposition in the derivation or an assumption or supposition. And even more strongly than this, '. . . no evaluation of *if p then q* is possible when *p* is false (and its falsity is not set aside by a deliberately counterfactual supposition)' (Braine and O'Brien, 1991, p. 188).

These restrictions have the ironic result that Braine and O'Brien's system cannot properly be called a mental or any other sort of *logic* at all. The restrictions clearly prevent 'if p & not-p then p' from being inferred as a logical truth, even though 'if p & not-p then p' is a substitution instance of 'if p & q then p'. Thus, in their 'logic', validity is not a matter of logical form. Not even and-elimination is formally valid in their system, as p will not validly follow there from 'p & not-p' as an assumption. And of course MP also fails to be formally valid in their system, as p does not follow from 'if p & not-p then p' and 'p & not-p' as premises. By Braine and O'Brien's rules, all derivations have to stop given a contradiction or inconsistency.

Suppose we know that a student, Jane, has always done well in the past in her exams when she has revised for them, but we have no idea whether she is revising for her next exam. Consider the conditional statement that, if Jane is revising for her next exam, then she will pass it. According to what Braine and O'Brien said, in the quotation above, no evaluation of this conditional is possible if Jane is not, in fact, revising for her next exam. They will not even allow us to infer that, if Jane is revising for her next exam, then she is revising for her next exam. Apparently, if p happens to be false, whether we know it or not, we cannot infer 'if p then p' 'validly' in their sense. By what Braine and O'Brien strictly say, 'validity' in their sense depends on the state of the actual world independent of our knowledge. This means that their notion of 'validity' could not be farther from validity in the proper logical sense. Braine and O'Brien would allow us to make some counterfactual assertions in the face of what we assume or know to be false, or what happens to be false. However, this does not help them in what they said about indicative conditionals, and we will find problems with their account of counterfactuals in Chapter 7.

Braine and O'Brien (1991) discussed the two paradoxes of the material conditional, which we described in Chapter 2. They gave these as:

4.4   Given q, one can infer if p then q
4.5   Given not-p, one can infer if p then q.

They naturally rejected 4.5 as invalid for their conditional. To derive 'if p then q' from not-p by conditional proof would require assuming both p and not-p, and that is of course blocked by Braine and O'Brien's rule that derivations must stop given an inconsistency. On the other hand, they accepted 4.4 as valid, remarking about their 'model' of the logic of the conditional (p. 197):

Instances of [4.4] are pragmatically strange, because if one already knows that *q*, what is the point of wondering whether *if p then q*? However, if one sets aside the strangeness and considers the issue on its merits for some specific *p* and *q* (i.e., whether *If p then q* is true, false, or undecidable, given that *q* is true), the model predicts that the answer has to be 'true'.

Braine and O'Brien erroneously claimed (p. 197) that inference form 4.4 is also valid in the conditional logic of Stalnaker (1968). They later repeated (p. 199) this incorrect interpretation of Stalnaker, claiming that they agreed with him that 4.4 is valid. But as we saw in Chapter 2, the fact that inferences of the form 4.4 are invalid follows immediately in Stalnaker type semantics for the conditional. Going back to our example there, we cannot validly infer that, if it rains, then plants will die, merely because the plants will die in a drought in the actual state of affairs. In the closest possibility in which it rains, the plants will live.

Like Stalnaker, correctly understood, we cannot see an advantage in a halfway paradoxical house that imagines 4.4 to be valid, and only 4.5 invalid, for the ordinary indicative conditional. More deeply, we cannot see value in merely listing a set of supposedly valid inferences for this conditional and then claiming that nothing else is required as semantics. Braine and O'Brien claim that their conditional has a semantics and thus a content. They would deny that we are expressing a conditional degree of belief when we use their conditional, but what really is its content? If we use a conditional 'if p then q' in their sense, what assertion are we making about the actual world or a possible states of affairs? We are not asserting 'not-p or q', for we would then be using the material conditional. We are not stating that q is true in the closest possibility in which p is true, for we would then be using a Stalnaker conditional. We cannot just stipulate that we are asserting whatever content is determined by Braine and O'Brien's rules. This is worse than claiming that intelligence is whatever IQ tests measure. There is far more to be said psychologically about intelligence than that one has done well in an IQ test. There is also far more to be said psychologically about the content of an indicative conditional than that it satisfies Braine and O'Brien's supposedly valid inference rules.

One of our aims in this book is to connect the effective use of conditionals with hypothetical thought. Consider the application of hypothetical thought to decision making. Suppose that we want to make a decision about whether to go for a walk. The walk will be less pleasant in the rain, and we notice that the barometer is failing. We have some confidence that,

4.6   If the barometer is falling, then it will rain

Our confidence in 4.6 is a mental state that is of importance in our decision making about whether to go for a walk. This confidence must in turn be related in some way to our confidence in possible states of the barometer and the weather, at the very least to the 'b & r', TT, and to the 'b & not-r', TF, possibilities. The relative probabilities of these TT and TF possibilities determine the conditional probability, $P(r|b)$, and that is directly relevant to decision making. But Braine and O'Brien did not have any account of how we make a judgement about what confidence to have in 4.6. They denied that they had to explain the content of 4.6 in terms of possible states of affairs. They imagined that they did not have to supply such a proper semantics for 4.6, yet that makes it totally unclear how 4.6 can be used in decision making.

Braine and O'Brien, unlike Rips and Marcus, did not see the need for a procedure, such as the Ramsey test, which is more general than conditional proof. Their version of conditional proof, like classical if-introduction, establishes that a conclusion is valid, i.e. necessary given some premises. It cannot show, unlike the Ramsey test, that a conclusion is probable, to a lower degree than necessary, given premises. Braine and O'Brien could not simply solve this problem extending their system and having rules for inferring conclusions such as:

4.7   If the barometer is falling (b), then it will probably rain (r)

Note that 4.7 would be another Braine and O'Brien conditional, and we could have more or less confidence in it. We must distinguish our confidence in a whole conditional of the form 'if p then probably q' from the degree of probability explicitly or implicitly expressed in its consequent, 'probably q'. For another example, we might have very high confidence that a coin is fair. In other words, our confidence might be very high that, if the coin is spun, then the probability that it will come up heads is 0.5. Braine and O'Brien did not give us any account of the probability of a whole conditional. They could not do this until they had semantics for their conditional in terms of the possible states of affairs that are fundamental to probability judgement. We must know what a conditional asserts about possible states of affairs, which have some relative probability, before we can know what confidence to have in it.

Rips and Marcus (1977) much more clearly distinguished questions of valid inference from what justifies, and gives us confidence in, a contingent conditional. (See also Rips, 1994, on how to formulate precisely a general mental logic.) As we have explained above, Rips and Marcus proposed that our confidence in a contingent proposition, P (if p then q), equals the conditional probability, P(q|p). Unfortunately, as we also explained above, this claimed identity does not follow from Stalnaker's semantics for the conditional. How the semantics of the conditional and conditional probability are connected with each other is a deep problem. But the experimental evidence is strong that the probability of a conditional is in general represented by the conditional probability (Evans *et al.*, 2003b; Oberauer and Wilhelm, 2003; Over and Evans, 2003). Our view is that, to make progress on a theory of conditional reasoning, and its application to decision making, we must investigate further people's representation of P(if p then q) as P(q|p).

## The Johnson-Laird and Byrne conditionals

Rips and Marcus (1977) and Braine and O'Brien (1991) give accounts of the ordinary indicative conditional as part of their theories of mental natural deduction systems, with introduction and elimination inference rules for the conditional and other sentential connectives (see also Rips, 1994; Braine and O'Brien, 1998a,b). In apparent contrast, Johnson-Laird and Byrne (1991) took a model theoretic approach to the study of the conditional and inference in general. They tried to explain human inference by proposing that people manipulate 'mental models'. A mental model is a human representation of a possible state of affairs that has, to some extent, the same structure as that state of affairs (Johnson-Laird and Byrne, 1991, p. 38).

In this section, we will discuss the Johnson-Laird and Byrne (1991, 2002) theories at length, examining in detail a number of their proposals and implicit features. We do this because the mental model theory is by far the most popular in the psychological study of conditionals and constitutes in effect the dominant paradigm for this field of study.

The mental models research programme has inspired many experimental studies and the study of new phenomenon, but it has a number of serious limitations that must be overcome if understanding of this important topic is to develop. As we will show, the (2002) theory is only well formulated with regard to what these authors term 'basic conditionals', which are treated as material conditionals in a version of T1. Non-basic conditionals are those where background and contextual knowledge modify the mental models in some way by a mechanism Johnson-Laird and Byrne term *pragmatic modulation*. It is not entirely clear how this works, but all of their detailed examples utilize the extensional approach in which the effect of knowledge is to add or subtract from a set of 16 logical possibilities. Even so, their influential theory deserves extended discussion.

Let us start by contrasting the model theory with natural deduction systems based on inference rules. Consider how the different approaches treat conjunction, 'p & q', and disjunction, 'p or q'. In a mental natural deduction system, there are introduction and elimination rules for conjunction and disjunction. For example, the and-elimination rule states that we may validly infer p (and also validly infer q) from 'p & q' as a premise, and the or-introduction rule states that we may validly infer 'p or q' from p (or from q) as a premise. In a mental models approach, conjunctions and disjunctions are broken down into their components, which represent the possible states of affairs that make the conjunctions and disjunctions true. The mental representations 'model' the possible states of affairs, which correspond to rows of the truth tables for the conjunctions and disjunctions, as can be seen in Table 4.1. Return to the example we used above of a conjunction:

We go for a walk and sprain our ankle

The mental models for this are:

We go for a walk    We sprain our ankle

There was also our example of a disjunction above:

We go for a walk or have a swim

The mental models for this are:

We go for a walk
                            We have a swim

Here each separate line represents a new possibility, and a third line can explicitly represent the possibility in which we both go for a swim and have a walk. Clearly, the procedure for constructing the mental models of conjunction and disjunction is essentially the same as the process that Rips and Marcus (1977) described for generating 'suppositions'. Johnson-Laird and Byrne, however, had a very different account of the conditional.

Johnson-Laird and Byrne (1991) gave a pure T1 account of the ordinary indicative conditional that presupposed the correctness of a truth table analysis of it (pp. 73–74). That is, Johnson-Laird and Byrne interpreted the ordinary indicative conditional, when the mental models for it are fully represented, or explicitly 'fleshed out', as equivalent to the truth functional, material conditional of formal logic. This analysis implies that 'if p then q' has the same explicit mental models as the disjunction 'not-p or q'. In more technical terms, the three fully explicit mental models for the material conditional represent the three possible states in which this conditional is true. These are the 'p & q', or TT, state, the 'not-p & q', or FT, state, and the 'not-p & not-q', or FF, state. And of course the confidence we should have in this conditional is clearly defined, as its probability is equal to P(p & q) + P(not-p & q) + P(not-p & not-q).

Johnson-Laird and Byrne referred to the mental models that represent these three possible states in this way:

```
p    q
¬p   q
¬p   ¬q
```

The three lines above correspond to the three rows of the truth table in which the material conditional is true. It is important to note, however, two important psychological proposals in their theory which diverge from a standard truth table. First, they propose that in accordance with a 'principle of truth', people construct only mental models to represent what is true and not what is false. In the above representation, the false case of 'p & not-q' is indicated only implicitly by its omission from the list of true possibilities. The principle of truth has been an important assumption in the current mental model theory. It has been used to try to explain the existence of several kinds of 'illusory inference' that people are claimed to make (Johnson-Laird and Savary, 1999; Goldvarg and Johnson-Laird, 2000; Yang and Johnson-Laird, 2000). The second important difference is that Johnson-Laird and Byrne (1991, p. 47) proposed that many people will not initially have fully explicit models for the conditional, but will only have shortened, initial models for it represented in this way:

```
p   q
. . .
```

The three dots in the above stand for implicit models of the false antecedent states, FT and FF. Johnson-Laird (1995) calls these dots a 'mental footnote' that there are implicit models in which not-p holds. See Table 4.1 for a summary.

Johnson-Laird and Byrne (1991, p. 67) also used square brackets in initial mental models of the conditional like this:

```
[p]   q
. . .
```

They stated that the square brackets indicate that p does not occur in any of the implicit mental models, denoted by the three dots. This use of square brackets has been heavily criticized by linguists, logicians, and other psychologists (Andrews, 1993; Hodges, 1993; Braine, 1993). Johnson-Laird and Byrne did not give precise formal rules for these brackets. It is hard to see what they mean, unless [p] q is a notational variant of 'if p then q', and unless the rules for manipulating the brackets are notational variants of MP and other inferences in a mental logic. In the most recent mental model theory of conditionals, Johnson-Laird and Byrne (2002), the brackets rightly make no appearance.

The system of fully explicit mental models for the indicative conditional, and conjunction and disjunction, is essentially equivalent to representing what are called disjunctive normal forms (Jeffrey, 1967). One can derive this extensional normal form for a truth functional proposition from the rows of its truth table in which it is true. Each of these rows describes a distinct possibility, and the proposition is true if and only if one of these distinct possibilities holds. To form the disjunctive normal form, we write out each of these rows as a conjunction, and then write out a disjunction of each of these conjunctions. The disjunction asserts that one of the conjunctions holds, and each conjunction asserts that one of the distinct possibilities hold.

For example, the disjunctive normal form for the material conditional states that either the TT, FT, or FF possible state holds:

(p & q) or (not-p & q) or (not-p & not-q)

The material conditional is logically equivalent to the above disjunctive normal form. The fully explicit mental models for the conditional represent, in effect, this normal form. Each distinct mental model, which Johnson-Laird and Byrne represent as a separate row, corresponds to one of the conjunctions in the above, and the full list of mental models corresponds to the overall disjunction. Johnson-Laird and Byrne have, by their convention, a nice visual way of representing the disjunctive normal forms. This is taken by them to correspond to some natural representation in people's mental states, the mental models themselves.

There are a total number of 16 disjunctive normal forms for two atomic propositions, p and q, which do not contain other propositions as components. Johnson-Laird and his collaborators try to do a great deal with these 16 extensional forms, which represent 16 logically possible representations. They use only these representations for the mental model theory of indicative conditionals. Even more strikingly, they only use these representations for counterfactual conditionals and to try to explain naive probability and causal judgement and reasoning (Johnson-Laird and Byrne, 1991, 2002; Johnson-Laird *et al.*, 1999). In information theoretic terms, one can distinguish their 16 representations with four bits of information, and so we call this approach the four-bit semantic device (Evans *et al.*, 2003c). In so far as the 16 disjunctive normal forms are extensional, truth-functional propositions, their approach appears to be completed extensional.

Johnson-Laird (personal communication) has argued that their mechanism of pragmatic modulation is not limited by the four-bit device as models can also represent probabilities and directional links. We deal with their (logical) representation of probability below. There is a passing reference in Johnson-Laird and Byrne (2002, p. 667) to directionality, when they say, 'Another factor that should affect conditional inferences is the so-called figural effect, which is the tendency for reasoners to draw conclusions interrelating items in the same order in which items entered working memory.' They also refer to apparently non-extensional concepts in commenting on belief bias, when they say that participants may resist inferences from false premises or to false conclusions. The difficulty we see is that they appear to have no explanatory mechanism other than the four-bit device to deal with such pragmatic influences. There is certainly no specification of a separate reasoning mechanism for non-basic conditionals other than reasoning by possibilities (and it would be strange if there were). We note also that their detailed accounts and supporting experiments deal only with extensional explanations of inference. For example, MT will be more often made when it requires identification of a mental model (true possibility) that is compatible rather than incompatible with prior belief (their Experiment 4). Though Johnson-Laird and Byrne state that pragmatic influences affect their mental models, we see no basis for crediting the theory with a mechanism beyond the four-bit device of extensional semantics. This means that contextual influences can only add or subtract possibilities.

The 16 disjunctive normal forms represent 16 logically possible representations, but Johnson-Laird and his collaborators do not use these possibilities to formulate an intensional semantics for the conditional. They make no attempt to define relative closeness between mental models or the represented possibilities. Leading philosophical logicians (Stalnaker, 1968; Lewis, 1973) and psychologists (Rips and Marcus, 1977) have intensional semantics for ordinary conditionals, including accounts of how one possibility can be relatively similar to, or relatively 'close to', another possibility. In judgement and decision making and social psychology, ordinary people have been found to think of some possibilities as closer than others (Kahneman and Miller, 1986; Roese, 1997, 2004; Teigen, 1998, 2004). However, Johnson-Laird and his collaborators make no use of this work, philosophical or psychological. They ambitiously try to explain a significant type of naive probability judgement by supposing that people, as a default, think of all these logical possibilities as having equal probabilities (Johnson-Laird *et al.*, 1999). They agree, of course, that people judge some mental models and the represented possibilities to be more probable than others. However, they take no account of psychological closeness, in spite of all the work there is in judgement and decision making on how closeness and probability judgements are interconnected. Their work is limited, for all its promise. Little can be done in explaining conditional reasoning, or probability judgement and causal reasoning, in their extensional approach (Over, 2004a,b).

We believe, in spite of the above comments, that Johnson-Laird and Byrne (1991, 2002) are thoroughly committed to the four-bit device. It follows that they are committed to the truth table analysis of the ordinary indicative conditional, making that, in their model theory, equivalent to the truth-functional conditional. They used (1991, p. 7) this example:

4.8   If Arthur is in Edinburgh, then Carol is in Glasgow

They asked whether 4.8 is true or false when Arthur is not in Edinburgh. They answered (p. 7), 'It can hardly be false, and so, since the propositional calculus allows only truth or falsity it must be true.'

The 'propositional calculus' is classical extensional propositional logic and its conditional is the material conditional. Johnson-Laird and Bryne are committed to this logic in the mental model account they give of negation, conjunction, disjunction, and the indicative conditional. This follows from their explicit statements, such as the one just quoted, and the relation between their mental models and disjunctive normal forms.

Johnson-Laird and Byrne (1991, p. 8) held that the use of an ordinary conditional can 'suggest' a relation, such as one of causation, between the antecedent and consequent. But they were clear that this relation could not enter into the semantic analysis of the conditional and make it true or, if absent, false. If the relation does not exist, the conditional assertion may be pragmatically inappropriate, but it is still true whenever its antecedent is false.

As an example, we might assert 4.8 because we believe that there is a relation between Arthur's being in Edinburgh and Carol's being in Glasgow. We might think that Arthur and Carol have joint work to talk about at the same time at Edinburgh University and Glasgow University. We infer that they will come to an agreement about which one will go where, and it is on this basis that we believe 4.8. But suppose we are wrong, and they do not have to present their work at the same time in different places, and in addition, Arthur is not in Edinburgh. Our assertion of 4.8 might be pragmatically misleading, but Johnson-Laird and Byrne (1991) would have argued that 4.8 is true in this kind of case. By claiming that 4.8 'can hardly be false' when the antecedent of 4.8 is false, Johnson-Laird and Byrne showed how deeply they were committed to the truth table analysis of the ordinary indicative conditional. Other theorists would of course take a very different view. For Stalnaker (1968) and Rips and Marcus (1977), 4.8 is false in the case described. We were wrong about how Arthur and Carol had joint work to present at the same time in the different places, and there is no real reason for Carol to be in Glasgow given the supposition that Arthur is in Edinburgh. Thus in the closest possible state of affairs in which Arthur is in Edinburgh, Carol is not in Glasgow, and that makes 4.8 false.

Johnson-Laird and Byrne (1991) had to argue, given their mental model theory, that the paradoxes of the material conditional are also valid inferences for the ordinary indicative conditional. They correctly pointed out (p. 74):

> Theorists therefore face a choice: to abandon the truth table analysis of conditionals (even if it is supplemented by inferences based on general knowledge), or to accept these apparently paradoxical deductions and to explain why they seem improper. We shall embrace the second alternative.

As we saw in Chapter 2, a conditional is a truth functional, and so material, if and only if the P1 and P2 paradoxes are valid for it. This fully justifies Johnson-Laird and Byrne's dilemma argument above, but by adopting the second alternative, they must explain why the supposedly 'valid' paradoxes seem to be paradoxical.

How Johnson-Laird and Byrne (1991) tried to explain the paradoxes (pp. 74–75) away is best illustrated by their example of an ordinary conditional with a negated antecedent:

If Shakespeare didn't write the sonnets then Bacon did

As they pointed out, their account of the ordinary conditional as the material conditional implies that this example is logically equivalent to:

Shakespeare wrote the sonnets or Bacon did

They went on to argue that both of the following inferences are valid and only seem paradoxical or odd for the same reason:

Shakespeare wrote the sonnets
Therefore, if Shakespeare didn't write the sonnets then Bacon did
Shakespeare wrote the sonnets
Therefore, Shakespeare wrote the sonnets or Bacon did

Their point was that both of these inferences 'throw semantic information away', in that the single premise is more informative than the conclusion, and doing that violates 'one of the fundamental constraints on human deductive competence' (p. 74). This argument that the paradoxes are only an apparent problem, for what we call the T1 family, has been a constant theme of the theory (Johnson-Laird, 1995; Johnson-Laird and Byrne, 2002).

However, the argument loses whatever force it has when uncertain premises are taken into account. Few premises in ordinary reasoning can be just assumed true. We cannot usually, with complete safety, assume that a proposition is true when it has some probability that is less than certainty. We must take account of that degree of probability in our assertions and our inferences, if we are not to mislead ourselves as well as other people. Until relatively recently, psychological theories of reasoning were severely limited by covering only inferences from assumptions and not from premises uncertain to some degree. This limitation is being overcome in response to experiments where valid inferences are suppressed by uncertainty in the premises (Byrne, 1991; Politzer and Braine, 1991; Stevenson and Over, 1995, 2001; Politzer and Bourmand, 2002). We discuss this phenomenon in Chapter 6. Some inferences should be suppressed by uncertainty in the premises, so that one can avoid believing or asserting an improbable conclusion, but other inferences should be encouraged by uncertainty in the premises, particularly those in which information is 'thrown away'.

Suppose that we recall hearing at school about some dispute over whether Shakespeare, Bacon, or someone else wrote the sonnets. Perhaps we took no interest in the sonnets, Shakespeare, or Bacon after school. We think, but we are not sure, that our teacher was quite negative about the claim that Shakespeare did not write the sonnets, and we have fair, but not absolute, trust in that teacher. We believe to some degree short of certainty that Shakespeare wrote the sonnets. But when we come to infer something to assert from this belief, we may well prefer to infer and to assert that Shakespeare wrote the sonnets or Bacon did. That is the safer and more reliable inference and assertion, and less likely to get us in trouble for misleading our hearer than if we boldly asserted that Shakespeare wrote the sonnets.

There are sometimes rational grounds for inferring a disjunction from one of its disjuncts. This supposedly 'throws semantic information away' and that is ruled out in mental model theory by one of its 'fundamental constraints'. But that constraint is there only because mental model theory is limited to inference from assumptions and does not cover uncertain premises. When a premise is uncertain, it is sometimes rational to lose semantic information in an inference, as that leads to a more probable conclusion than the premise. There is nothing at all 'paradoxical' about playing safe and inferring a more probable disjunction from one of its less probable disjuncts. However, it is just plain wrong to assert, and even more obviously to *believe*, an example such as RD in Chapter 2, that the plants will die if it rains, merely because there has been a drought and it is improbable that it will rain. (See Jackson, 1987, for a stronger pragmatic justification of the paradoxes than Johnson-Laird and Byrne's, and Edgington, 1995, for criticism of Jackson.)

Johnson-Laird and his collaborators have maintained until recently their commitment to the truth table analysis of the ordinary indicative conditional. They have claimed that people perform 'illusory' inferences in conditional reasoning if these are not in line with the logic of the material conditional (Johnson-Laird and Savary, 1999). People also supposedly make mistakes in their probabilistic reasoning about conditionals if they do follow the logic of the material conditional (Johnson-Laird *et al.*, 1999). However, in their most recent paper on conditionals, Johnson-Laird and Byrne (2002) have a different position, which is very unclear in some respects.

Johnson-Laird and Byrne (2002) begin by proposing a mental model of theory of what they call *basic* conditionals, which are defined (p. 648) as having:

> . . . neutral content that is an independent as possible from context and background knowledge, and which have an antecedent and consequent that are semantically independent apart from their occurrence in the same conditional.

An example of a basic conditional would be, 'if there is a circle then there is a triangle', when this is asserted about some shapes on cards that we are unfamiliar with. The conditionals about Shakespeare and Bacon above would not be basic, at least for most of us, who have background knowledge about these men. The conditional 4.8 would not be basic either, if we did know that Arthur and Carol were collaborators and had to give the same talk at Edinburgh University and Glasgow University at the same time. It is unclear, however, whether simply having background knowledge about Edinburgh and Glasgow would make 4.8 non-basic.

What Johnson-Laird and Byrne (2002) say about truth and falsity is more deeply unclear. They describe (p. 653) how each mental model 'corresponds to a row in a truth table' and state a fundamental 'principle of truth':

> Each mental model of a set of assertions represents a possibility given the truth of the assertions, and each mental model represents a clause in these assertions only when it is true in that possibility.

They go on to claim (p. 653) that '. . . possibilities are psychological basic, not truth values'. But this cannot be so when there is a fundamental principle of truth. Moreover, Johnson-Laird and Byrne use the term 'satisfied' in a way that can only make it a synonym of 'true'. For example, they remark (p. 673), 'Basic conditionals have mental models representing the possibilities in which their antecedents are satisfied, but only implicit models for the possibilities in which their antecedents are not satisfied.' They cannot do their semantics at all without talking about when assertions and their component propositions are, or are not, 'satisfied' in possible states of affairs. This is to give the truth conditions of the assertions, but to use the words 'satisfied' and 'not satisfied' in the semantics. It is much clearer to use 'true' and 'false' when talking about truth conditions, and we will continue to do so in our attempt to interpret Johnson-Laird and Byrne (2002).

Johnson-Laird and Byrne (2002) present a model theory of basic conditionals that is the same as the model theory that was proposed for all ordinary indicative conditionals in Johnson-Laird and Byrne (1991), as described above and shown in Table 4.1. The fully explicit mental models for basic conditionals correspond to the disjunctive normal form of the material conditional. Basic conditionals must be truth functional conditionals. Johnson-Laird and Byrne (2002, pp. 651–652) also continue to defend the validity of the paradoxes in the way we have explained, and criticized, above. This is at best a misleading defence of the paradoxes, in that they do not say that it is restricted to basic conditionals. They later claim (p. 673):

> Conditionals are not truth functional. Nor, in our view, are any other sentential connectives in natural language.

This statement is again misleading, if not inconsistent with what they claim about the paradoxes. They do not point out that it can only apply to non-basic conditionals, those subject to pragmatic modulation. As we proved in Chapter 2, and have noted above in this chapter, accepting that the paradoxes are valid for a conditional is logically equivalent to holding that that conditional is truth-functional. Johnson-Laird and Byrne's position must be that the paradoxes are valid for basic conditionals and so these are truth functional, but invalid for non-basic conditionals and so these are not truth functional. These two Johnson-Laird and Byrne conditionals, the former truth functional and the latter not, must have different mental models. But there are not clear mental models for non-basic conditionals, and it is obscure what determines which inferences are valid for them. Johnson-Laird and Byrne do not introduce an intensional semantics (Chapter 2) for non-basic conditionals and, in danger of outright inconsistency, appear to presuppose that their extensional semantics can somehow account for these non-truth functional conditionals.

Johnson-Laird and Byrne (2002) go so far as to claim, in the quote just given above, that no connective in natural language is truth functional. (The exception is that negation is truth functional; Johnson-Laird, personal communication.) This claim is also a significant departure from the account given of propositional reasoning generally in Johnson-Laird and Byrne (1991). The new position is the result of reflecting (Johnson-Laird, personal communication) on examples discussed by Strawson (1952). Consider Strawson's example of a use of 'and' that is not supposed to be truth functional, 'They got married and had a child.' If this statement were truth functional, it would be logically equivalent to, 'They had a child and got married.' Of course, we might blame the lack of truth functionality here on an implicit tense operator. We might also try to interpret such examples as ones in which the speaker makes a pragmatic suggestion about a relation in time (Grice, 1989). If we look at the example in either of these ways, we would not conclude that 'and' itself has some uses that are not truth functional. But if there are such uses, Johnson-Laird and Byrne (2002) do not give us a mental model theory of them, do not tell us their logic, and do not say what their probabilities can be. Obviously, if 'p & q' is not logically equivalent to 'q & p', standard propositional logic does not apply, nor does standard probability theory, in which P (p & q) and P (q & p) are necessarily equal. What is the mental model of a 'p & q' not logically equivalent to 'q & p'? Johnson-Laird and Byrne cannot answer this question with the mental model for conjunction in Table 4.1. That is only for truth functional conjunction, and yet they proffer nothing else. Similar points apply to their non-basic, non-truth functional indicative conditionals.

**Table 4.1.** Mental models for conjunction, disjunction, and the material conditional

| Connective | Initial mental models | | Fully explicit models | |
|---|---|---|---|---|
| p & q | p | q | p | q |
| p or q | p | | p | ¬q |
| | | q | ¬p | q |
| | p | q | p | q |
| If p then q | p | q | p | q |
| | . . . | | ¬p | q |
| | | | ¬p | ¬q |

Johnson-Laird and Byrne's mental model theory of conditional reasoning only applies to truth functional 'basic conditionals'. Consider their account of the difference between MP and MT. There is a higher rate of endorsement of MP as valid than of MT as valid (see Chapter 3). Johnson-Laird and Byrne (1991) explained this in terms of the difference between the initial models of the indicative conditional and the fully explicit models of this conditional. People will judge MP to be valid on the basis of the initial models. But according to Johnson-Laird and Byrne, people must make the models fully explicit, as in the final column of Table 4.1, to grasp the validity of MT. To make the models fully explicit is to generate, in effect, the disjunctive normal form for the material conditional. By doing this, one makes the two not-p possibilities, FT and FF, explicit. To respond on this basis that MT is valid, people must conclude that the FF possibility makes the conditional automatically true. But it would be a fallacy to conclude that the FF possibility makes a non-truth functional conditional automatically true. For this reason, Johnson-Laird and Byrne's account of MT, and the difference between MP and MT, applies only to basic, truth functional conditionals. They have no account of MT for non-basic, non-truth functional conditionals. There is no entry for these intensional indicative conditionals in Table 4.1, because Johnson-Laird and Byrne do not present mental models of them.

Johnson-Laird and Byrne's mental model theory is primarily extensional. It is only intensional in covering *logical* possibility and *logical* probability. Not much can be explained about human reasoning and judgement with these logical notions (Over, 2004a,b). Johnson-Laird and Byrne should develop a clearer and more plausible mental model theory of at least non-basic conditionals by taking one of two approaches. They could introduce a relation of 'closeness' between mental models of logically possibilities, making use of the psychology of closeness judgements studied in judgement and decision making and social psychology. In this way, they could place their theory clearly in T2, turning it into a psychological version of Stalnaker's analysis. Alternatively, they could adopt a T3 approach and try to explain, in mental model terms, how conditionals are evaluated as conditional probabilities. In either case, they should move beyond their limited logical concept of probability, with its default assumption that all mental models are equally probable in this sense. They could then explain why some mental models are more probable than other mental models, because of relative frequency information or the use of heuristics, or because of some other kind of evidence or inference.

Johnson-Laird and Byrne (2002) do briefly refer (p. 652) to Stalnaker's extension of the Ramsey test. They apparently do not want to make use of it because they associate it with what Stalnaker and other philosophers have said about counterfactuals and 'possible worlds'. We will cover counterfactuals and 'possible world' semantics in Chapter 7. But Stalnaker's extension of the Ramsey test is independent of his semantic analysis of conditionals in terms of possible states of the world. In the first place, his extended Ramsey test is psychological in nature and can be developed without objectionable philosophy or metaphysics. His extension is about minimal changes that people are to make to assumptions or beliefs to remain consistent when they think about the antecedent of a conditional hypothetically. There is only psychology in that and not philosophy or metaphysics. In the second place, as we saw in Chapter 2, Lewis (1976) showed that Stalnaker's semantic analysis does not, in fact, precisely reflect his extension of the Ramsey test. His extension still results in a conditional probability judgement, but that is not identical, in general with the probability of a Stalnaker conditional.

Johnson-Laird and Byrne (2002) also express doubt (pp. 650–651) about the relation between an ordinary conditional and conditional probability. But the evidence is strong that this relation exists especially for non-basic conditionals (Over and Evans, 2003, and see Chapter 8). Johnson-Laird and his collaborators have no theory of the probability of a non-basic conditional. At present, the content of the Johnson-Laird and Byrne non-basic conditional is as indefinite as that of the Braine and O'Brien conditional, and we have no more idea of how the one can have a probability than how the other can. In fact, Johnson-Laird and Byrne are in an even weaker position, as they leave us in the dark about the logic and probabilities of non-truth functional conjunctions and disjunctions.

## Can the mental model theory be fixed?

As indicated above, the mental model theory has formed the dominant paradigm in the psychology of reasoning for the past decade and more and many authors studying conditional reasoning have attempted to interpret their findings within the Johnson-Laird and Byrne (1991) theoretical framework. As also shown above, the revision presented by Johnson-Laird and Byrne (2002) has not significantly changed the earlier theory as far as basic conditionals are concerned. In essence, it is a T1 account, holding that the meaning of basic conditionals can be captured by the extension of logical possibilities (truth table cases) that are allowed, and all conditional inferences are a consequence of people's imagination and manipulation of these logical possibilities.

Several authors have, however, tried to extend or modify the model theory of conditionals in order to improve its ability to provide a plausible and accurate psychological account of the data, including the first author of this book. For example, during a period in which the first author of this book was trying to reconcile his heuristic-analytic theory of bias with the mental model account of deductive competence, he presented a critique and revision of the model theory of conditionals (Evans, 1993). We start by examining briefly the proposals made in that paper. Evans (1993) was concerned first about ambiguity in the theory with regard to the fleshing-out of initially incomplete mental models of the conditional and the conditions permitting people to draw each of the inferences MP, DA, AC, and MT. The revisions suggested in this regard did not, however, alter the essentially extensional nature of the account to which we have explicated strong objections above.

One part of Evans' (1993) proposals did, however, address a core objection in our current thinking. Johnson-Laird and Byrne (1991) had proposed an account of the differences in observed reasoning between 'if then' and 'only if' conditions, some of which were briefly discussed in Chapter 2. In particular, they wanted to account for why the MP inferences was made more often with the 'if then' form and the MT inference more often with the 'only if' form. In the spirit of the four-bit device they added an extra model to the *initial* representation of the latter, as follows:

If p then q   pq
　　　. . .

p only if q   pq
　　　¬p¬q
　　　. . .

Hence, with 'only if' conditionals, the FF case (not-p & not-q) was proposed to be explicitly represented as well as the TT case. Hence, MT is facilitated (the minor premise, not-q, eliminates the TT model, leaving FF) and MP has more competition. Evans (1993) pointed out, however, that the evidence in the literature also showed that people made more DA inferences with 'if then' and more AC inferences with 'only if'. In other words people make more forward inferences (MP, DA) with 'if then' conditionals and more backward inferences (AC, MT) with 'only if'. Later, Evans *et al.* (1999b, Experiment 3) were to show an analogous difference occurs with the Wason selection task, where people choose more p and not-p cards with 'if then' statements and more q and not-q cards with 'only if' statements.

The proposal that two models are used to represent the 'only if' conditional is inconsistent with the singularity principle of our own hypothetical thinking theory (Evans *et al.*, 2003c, see Chapter 9). In this, we propose that people only imagine one possible state of affairs at a time, albeit with a much richer model format than those proposed by Johnson-Laird and Byrne. In fact, there is clear empirical evidence that people do *not* add a model of FF as a true possibility to their initial representation of 'p only if q'. For example, in the study of Evans *et al.* (1996b), described in Chapter 3, people actually included slightly fewer, not more, FF cases in their arrays when asked to represent as true 'only if' conditional. Similarly, the hypothesis predicts that people will more often evaluate FF cases as true in a truth table task with 'only if' than 'if then' cases, whereas the data show if anything the reverse trend (Evans, 1975; Evans and Newstead, 1977).

Clearly the differences between 'if then' and 'only if' conditionals, taken in conjunction with the temporal context effects for 'if then' and 'only if' that we discussed in Chapter 3, simply cannot be explained by the extensional approach. There are simply no logical possibilities that can be added or subtracted by pragmatic modulation that will sanction MP without MT, or AC without DA. What Evans (1993) suggested then was that directionality be incorporated within the mental models themselves. For example, if a 'p&q' model is built from p to q then MP will be facilitated and AC inhibited and vice versa. This, however, implies a radical departure from the standard Johnson-Laird and Byrne account, even if they (2002) considered such a possibility *en passant* as we mentioned earlier. Such mental models would no longer be simply logical possibilities, so a different inferential mechanism would need to be defined. Of course, directionality—if it could be accommodated—is not enough. Models must at the very least be able to represent uncertainty and mental degrees of connection between propositions as we have already argued.

Other authors working with the mental models paradigm have tried to revise the theory, sometimes with specific add-ons to account for specific new findings (for example, Schroyens *et al.*, 2000, 2001). In fact, many discussion sections in the journal articles implicitly or explicitly revise the Johnson-Laird and Byrne account in some way. More systematically, Barrouillet and colleagues (for example, Barrouillet and Lecas, 1998) have taken up the idea of directional, or as they call it 'relational' mental models from Evans (1993) and incorporated that within their own mental model account. Barrouillet and Lecas (1998) go further, by disputing also Johnson-Laird and Byrne's use of negation markers within mental models. They state (p. 213), 'It is debatable whether negation signs are compatible with the general theoretical framework of mental models . . . Mental models are . . . representations of situations, whereas the existence of negation markers makes them representations of classes of situations'. This is a very interesting point, and of course the use of negation markers in models is one of the devices that renders them essentially equivalent to truth table cases, and the underlies the extensional mental logic that we believe their system comprises. These ideas are developed further in the developmental theory of conditional reasoning put forward by Markovits and Barrouillet (2002). This theory addresses the severe limitations of the Johnson-Laird and Byrne account of pragmatic influences.

Markovits and Barrouillet (2002), while presenting what they call a 'revised mental models' account, make a number of similar points to ourselves. They comment (p. 8) that 'The current formulation of the mental model theory relies explicitly on a truth-table like analysis of reasoning . . . Whilst this might be true for at least some educated adult reasoners . . . such a hypothesis implies that claims that are not always consistent with the available data on how children reason and in at least some cases appear inconsistent with data on adult reasoners.' They go on to argue that the development of reasoning must build upon general cognitive architectures available to young children and that '. . . children (and adults) have an understanding of "if then" that is inherently relational and brings to bear a fairly rich linguistic and pragmatic experience' (p. 8). As in our own theory, presented in Chapter 9, these authors assume that the development of pragmatic reasoning precedes that of deductive reasoning and has at least an implicit dual process account in which these processes compete in adults. This theory places emphasis upon the development of ability to retrieve relevant information from semantic memory and the way in which this influences the models people consider and the consequent inferences that they draw.

The answer to the question of whether the (Johnson-Laird and Byrne) mental model theory of conditionals can be fixed is to some extent a matter of definition. At the broadest level of conception, the theory is right to propose that understanding and reasoning with conditionals is based on the construction and manipulation of mental models representing possible states of the world. Our own approach to a theory of 'if' conforms to this broad specification (Chapter 9). However, the specific account of Johnson-Laird and Byrne is, we believe, fundamentally wrong for the reasons we have elaborated here and elsewhere (Evans *et al.*, 2003c; Over, 2004a,b). By defining mental models as truth table cases, endorsing by implication extensional proposition logic, restricting the representation of uncertainty and providing no credible mechanism for pragmatic influences the current theory is very far from providing an adequate account of the phenomena of conditionals.

## Conclusions

The earliest significant psychological theory of ordinary indicative conditionals is that of Rips and Marcus (1977), and this is still the best in a number of respects. It depends on a clear logic with a proper semantics, that of the T2 Stalnaker conditional. It relates the probability of the conditional to conditional probability, and proposes an interesting procedure, comparable with the Ramsey test, for evaluating conditionals. But it also has major problems. It presupposes that the probability of its T2 conditional is identical to the conditional probability, and its procedure for evaluating conditionals is unbounded.

Braine and O'Brien (1991) and Johnson-Laird and Byrne (1991, 2002), in spite of their many differences, share some considerable problems. Both try to modify existing logical systems to make them plausible as bounded mental systems. Braine and O'Brien modify the rules of natural deduction, but leave the resulting conditional without a proper semantics and with no clear content. Johnson-Laird and Byrne make use, in effect, of the disjunctive normal form as the full mental models of their 'basic conditional'. They modify this form by having initial mental models for this conditional that are easy to construct. They do as much as they can with logical possibility and logical probability. However, most ordinary indicative conditionals are non-basic. It is obscure to our reading what Johnson-Laird and Byrne's mental model theory is of these non-truth functional conditionals. It cannot be in the T1 family, but that is all that we are able to say with any confidence.

# 5 Hypothetical thinking with 'if': the case of the selection task

Peter Wason established the modern study of the psychology of reasoning with a set of highly innovative and insightful papers published in the 1960s and 1970s. He invented three reasoning tasks that psychologists have investigated ever since: the '2, 4, 6' problem (Wason, 1960)—a test of inductive reasoning, the THOG problem (Wason and Brooks, 1979)—a rather convoluted test of disjunctive reasoning, and the four-card selection task (Wason, 1966), on conditional reasoning, that we will refer to simply as the selection task. Of these, the selection task has been by far the most intensively researched. In fact, there are far more papers written by psychologists on the selection task than any other problem used to study reasoning. However, there is a point of view—with which we have great sympathy—that too much research has been focused on this problem, which is not in any case a good method of measuring reasoning processes in general (Sperber *et al.*, 1995; Sperber and Girotto, 2002).

Some of the work on the selection task has been widely cited outside of the psychology of reasoning, but not always with great accuracy. For example, the belief that thematic material facilitates reasoning, following early research on the selection task discussed by Wason and Johnson-Laird (1972), achieved an almost mythical status in cognitive psychology for many years, despite rapidly emerging evidence in the field that the claim was greatly oversimplified. In recent years, we have similarly seen many references in cognitive science and philosophy to the effect that research on the selection task has provided good evidence for innate content or domain-specific reasoning mechanisms and support for a particular school of evolutionary psychology (Cosmides, 1989; Fiddick *et al.*, 2000). As we will see, this work and its interpretation was controversial from the start in the psychology of reasoning. Both the claims for innateness and domain specificity have been strongly contested by leading researchers in the field.

Although the selection task is not a good way of measuring deductive reasoning processes, for reasons we will explain, we believe that, correctly interpreted, psychological research on this task has proved very informative about how people understand and use 'if', and the nature of hypothetical thinking in general. Hence, understanding research on this task is most certainly relevant to the objectives of this book. As we have focused primarily on indicative conditionals so far, we will start with a discussion of the indicative selection task and defer consideration of the deontic selection task until later in the chapter.

## The indicative selection task

In Chapter 2, we reported psychological studies of reasoning with abstract indicative conditionals using two main methods: the truth table task, in which people are asked to judge the truth (and relevance) of truth table cases, and the conditional inference task in which people are asked to draw inferences from minor premises which assert or deny one of the component propositions of the conditional. The indicative selection task has been largely investigated by the same psychological researchers as a kind of third (and indeed the most popular) method of studying conditional reasoning. However, whether or not it should be classified as a conditional reasoning task at all is moot. Here is a typical example of the task. There are four cards lying on a table. Each is known to have a capital letter on one side and a single figure number on the other side. The following statement applies to these cards and may be true or false:

If there is an A on side of the card, then there is a 3 on the other side of the card

The four cards show on the their exposed sides the following symbols:

A     D     3     7

Your task is to choose those cards and only those cards that need to be turned over in order to decide whether the statement is true or false.

This task clearly is about an indicative conditional that purports to describe the four cards correctly. Logically, the statement can only be falsified by finding a card that has an A on one side and does not have a 3 on the other side. There are only two cards that, if turned, can lead to discovery of such case: the A and the 7. Wason (1966) therefore argued that A and 7 is the *correct* answer. More generally, for a conditional of the form 'if p then q', the p and not-q cards need to be chosen. Most (but not all) psychologists working the field since have accepted this normative analysis.

The reason this task is interesting is because most people get it wrong. They typically choose A or A and 3 (more generally p, or p & q). Solution rates vary according to the population studied but in typical undergraduate populations, it is not uncommon to find correct solution rates of 10% or less (Evans *et al.*, 1993). Those who do solve the task tend to have exceptionally high general intelligence scores (Stanovich and West, 1998a). The typical selection patterns on the abstract indicative selection task are robust with regard to many procedural variations (Wason and Johnson-Laird, 1972). These include precise wording of the statement (universal forms such as 'Every p is a q' make no difference), the use of binary or non-binary materials and the possibly ambiguous reference to the 'other sides' of the cards. For example, if the values are on the left and right sides of cards, one of which is masked, the results are essentially the same.

What does profoundly affect choices on this task is the presence of negative components. Readers will recall the account of the 'negations paradigm' in Chapter 2 and the consequent matching bias effect on the conditional truth table task. Matching bias on the selection task was first demonstrated by Evans and Lynch (1973) and has been replicated many times since. Let us refer to the four cards as True Antecedent (TA), False Antecedent (FA), True Consequent (TC), and False Consequent (FC). On the affirmative conditional these correspond to p, not-p, q, and not-q. So the typical choice of TA and TC is also the matching choice of p and q. Wason's original account of the error on the selection task was in terms of a *confirmation bias*. He suggested that people were trying to prove the conditional true rather than false. Hence, they chose the confirming cards TA and TC. As Evans and Lynch were the first to point out, this cannot be separated from the matching choices unless negations are introduced.

**Table 5.1.** Matching cases and selection task choices, illustrating the negations paradigm

|                     | TA    | FA    | TC    | FC    |
| ------------------- | ----- | ----- | ----- | ----- |
| If B then 3         | B     | E     | 3     | 2     |
| If p then q         | p     | not-p | q     | not-q |
| If G then not 6     | G     | K     | 4     | 6     |
| If p then not q     | p     | not-p | not-q | q     |
| If not R then 5     | M     | B     | 5     | 9     |
| If not p then q     | not-p | p     | q     | not-q |
| If not E then not 1 | K     | D     | 3     | 1     |
| If not p then not q | not-p | p     | not-q | Q     |

Table 5.1 illustrates the application of the negations paradigm to the selection task. For each logical case there are two cases across the four statement types that match (p or q) and two that mismatch (not-p or not-q). The effect is often described intuitively with regard to the facilitation of solution for the 'if p then not-q' form. For example, if the statement is

If there is a G on one side of the card then there is NOT a 6 on the other side of the card

and the cards shown are G, K, 4, and 6 then the matching cards G and 6 are also the logically correct choices TA and FC. This version of the selection task is very easy and almost everyone gets it right. Paradoxically, adding a negation has made the task much easier.

As Evans and Lynch observed, however, there is usually a significant matching bias effect on *all four cases*. That is, people choose more TA and FA cards that are p than not-p and more TC and FC cards that are q rather than not-q. Comparing logical cases averaged across the four types (that is with matching bias balanced) they found the following order of card choices: TA > TC = FC > FA. These two findings are illustrated graphically in Figure 5.1 taken from their data combined with those of three other similar experiments (see Evans *et al.*, 1993, p. 110).

These data provided strong support for the matching bias hypothesis and of course refuted Wason's original confirmation bias account, as he was quick to acknowledge at the time. Findings in other studies over the years have in some cases varied. For example, matching bias on the antecedent cards is not always strongly observed and some studies show an overall preference for TC over FC, leading some authors to suggest that there might be a confirmation bias as well as a matching bias. However, the findings shown in Figure 5.1 are typical and will form the basis for the following discussion.

## The heuristic-analytic account of the indicative selection task

Statistically speaking, the data of Figure 5.1 provide two findings that need to be explained: (1) a preference for true over false antecedent choices, and (2) a general matching bias effect. Evans (1984, 1989) offered an explanation in terms of the heuristic-analytic theory of reasoning biases an earlier form of the dual-process theory that

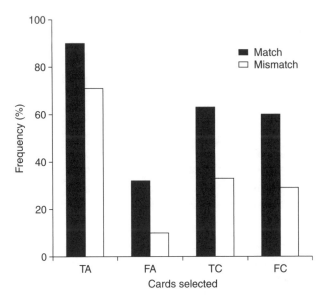

**Figure 5.1.** Percentage card selections over four experiments using the negation paradigm on the Wason selection task.

we described in Chapter 1. This theory proposed that preconscious heuristic (System 1) processes direct attention to problem information making it appear relevant. Subsequent analytic (System 2) processing is applied selectively to information represented as relevant. Specifically, Evans proposed the operation of two heuristics that we will term the if-heuristic and the matching-heuristic (also referred to as the not-heuristic by some authors). The if-heuristic was based on the linguistic usage of 'if' which, it was proposed, directed attention to the possibility of the antecedent being true.

As originally proposed the if-heuristic, accounting for the TA > FA effect, was an intuitive hypotheses that did little more than re-describe the data. However, consider the task from the viewpoint of the Ramsey test, and the two accounts of conditionals, T2 and T3, associated with this test (Chapter 2). You are asked to decide whether the statement is true or false. According to T3, a conditional only has a truth value in cases where the antecedent holds true, and thus T3 implies people will focus on true antecedent cases. (We have already noted support for T3 in the 'defective truth table' findings in Chapter 3, when as here abstract conditional statements are considered.) Arguably, in a T2 account, with its notion of the closest possibility in which the antecedent is true, a conditional would also focus attention on p and away from not-p. Hence, anyone seeking to apply the Ramsey test to the selection task must inspect the TA card and would see no relevance in the FA card. The suggestion that an if-heuristic operated as an unconscious and rapid process is fully consistent with this theory, as the Ramsey test (we hold) is so basic to an understanding of 'if' as to direct attention in this automated fashion (see Chapter 9).

What, however, of the matching-heuristic? In the original account of Evans (1984, 1989) the rationale was again in terms of linguistic function. Following Wason (1972) it was suggested that negatives were used to deny presuppositions rather than to assert new information. Hence, if your golf partner phones and you say immediately 'I cannot play golf today' the topic of this discourse is playing golf (denied) rather than an assertion of what you may be actually doing. By this reasoning the statement that 'there is an A on the card' and the statement that 'there is not an A on the card' are both *about* the A. Hence the letter or number referred to in the conditional forms the relevant topic of the discourse, whether negated or not. Thus matching cards seem more relevant than mismatching cards.

The heuristic-analytic account is widely interpreted as a proposal that there is no reasoning involved on the selection task and that its solution is purely heuristic. The first author (J. E.) feels that he is probably responsible for this belief even though he has not subscribed to it for a long time and not argued for it in any recent papers. There is a subtle problem here that is worth explaining. As illustrated above, the two heuristics seem to account entirely for the selection pattern shown in Figure 5.1. So apparently, analytic reasoning is not influencing card choices at all. Why should this be? The subtle distinction is this: analytic reasoning does occur on the selection task, but it *usually* does not affect the selections that are made. The reasons for this are to do with the peculiar nature of the task, and the main reason why we say that it should not be regarded as a good test of deductive reasoning.

There are two sources of evidence that show that people are reasoning on the abstract indicative selection task. The first lies in the use of verbal reports. One method involves asking people to provide a written justification for their card choices immediately after selection. The first authors to do this were Goodwin and Wason (1972). They found that reports revealed an appropriate level of insight into the task. For example, those choosing correctly referred to the need to falsify the conditional, whereas those showing typical errors referred only to verification. No one, however, described anything that sounded like matching bias and this experiment used only affirmative conditionals. In a subsequent experiment, Wason and Evans (1975) used both an affirmative (if p then q) and negative (if p then not q) statement. As noted earlier, matching bias leads people to make correct choices on the negative statement. Like Goodwin and Wason, they found an (apparently) appropriate level of insight in the protocols. Typical responses are illustrated below:

*Affirmative statement (if A then 3)*
Choose A because a 3 on the back would show the conditional to be true
Choose 3 because an A on the back would make it true

*Negative statement (if G then not 6)*
Choose B because a 6 on the back would prove the conditional false
Choose 6 because a B on the back would make it false

What is particularly interesting is that justifications for the TA card were given in terms of verification for an affirmative statement but falsification for the negative statement. Moreover, Wason and Evans found that people who attempted the negative problem before the affirmative problem showed an apparent level of insight that then disappeared, because they matched on both tasks. They hence argued that card choices were unconsciously determined and then rationalized by a separate conscious and verbal process. This led to an early version of the dual process theory of reasoning (see Chapter 1). In a later paper, they showed that people would justify any of four possible solutions to the task given to them as apparently correct by the experimenter (Evans and Wason, 1976).

At the time, it seemed that this rationalization was due to the retrospective method of verbal reporting, identified as problematic in a famous paper by Nisbett and Wilson (1977). However, subsequent use of 'think aloud' protocols—a method in which people talk continuously during performance of the task—produced quite similar findings (Evans, 1995). It was these later experiments plus those on card inspection times (described below) that provided convincing evidence that analytic reasoning was involved in the selection task. The protocols showed that people routinely considered the hidden sides of the cards before making their selections. However, they confined their attention (in general) not only to the matching cards but to the matching values on the *hidden* sides of those cards as well. That is why Wason and Evans found apparent insight into falsification on the negative statement problem. Given 'If there is a G on one side of the card, then there is not a 6 on the other side', people think about the G card and then consider the possibility of a 6 on the back (and vice versa). Thus they hypothesize the G6 combination and see this would be inconsistent with the conditional. This is precisely the kind of process envisaged in the heuristic-analytic theory: people do reason, but only about information selectively represented as relevant by the heuristic system.

The other main form of evidence that people are reasoning on the selection task is from card inspection times. Evans (1996) introduce a methodological innovation by asking participants in a computer displayed selection task to point the mouse at any card they were *thinking of selecting*, but only to click on the card when they were sure. The pointing time was accumulated, but only prior to a selection decision. As predicted, participants spent very little time pointing at cards that they did not go on to select, but spent far more time pointing at cards that were later selected. To account for this finding, you need both processes of the heuristic-analytic theory. According to the theory, cards are chosen only if attention is unconsciously directed to them by the heuristic process. This accounts for very low inspection rates of non-selected cards. However, why do people spend substantial amounts of time thinking about the cards they go on to select? The answer must surely be that they are seeking a justification via an analytic reasoning process.

The card inspection time methodology has been the subject of some debate in the psychological literature and there may be some problems with it (Roberts, 1998; Evans, 1998a). However, all of Evans's (1996) predictions have recently been confirmed using a superior methodology involving the tracking of eye movements (Ball *et al.*, 2003), which is not open to same objections. In addition, the original critic of this method, Max Roberts, has found recent evidence that is highly *favourable* to the heuristic-analytic theory (Roberts and Newton, 2002). In their first experiment they presented a 'change' task that they claimed to control for possible artefacts in the original method. In this task cards were shown as either selected or not selected and participants invited to change selections if they did not agree with them. Thus, for example, they had to point the mouse at cards to deselect them as well as to select them. In spite of this control, the predicted correlation between inspection time and selection was shown. In subsequent experiments, they used a rapid response selection task with a forced decision with 2 seconds. Compared with the standard selection task, there were significantly more matching responses for consequent cards. This suggests that even on the standard task, some modification of response is caused by analytic reasoning when there is time for it to occur. It also confirms the account of matching bias in terms of a rapid heuristic process, as otherwise the effect should have disappeared and been replaced with random responding on the rapid task.

A further recent finding suggests strongly that analytic reasoning may modify card choices on the selection task using normal populations. Feeney and Handley (2000; Handley *et al.*, 2002) demonstrated in several experiments that presentation of a second conditional with an alternative antecedent can affect card choices. Suppose the statement to be checked according to the usual instruction is:

If a card has a letter A on one side then it has the number 3 on the other side

The experimental group are *also* given the following information:

If a card has a letter L on one side then it has the number 3 on the other side

Compared with the control group, these participants show significantly fewer selections of the q card (or TC, as the experiments used affirmative conditionals only). It is well known that presenting an alternative antecedent like this can suppress the Affirmation of the Consequent fallacy on the conditional inference task (see Chapter 6). Hence, it would seem that the justification that people normally find for choosing the q card (i.e. that a p on the back would prove the conditional true) is similarly blocked by this manipulation.

So, if people are reasoning on the selection task, why do all but a small minority of high IQ participants get the problem wrong? And why does this reasoning have so little influence on the cards actually chosen in most experiments? There seem to be two reasons for this. First, people's attention is selectively and unconsciously determined. This explains why selection of the FC or not-q on the standard task is substantially lower than the rates of Modus Tollens normally observed with affirmative conditional rules (typically 60–75%). If you don't think about, you cannot select it. Second, the reason that people mostly justify the choice of cards they do think about is that they fail to appreciate the critical difference between verification and falsification. The protocols show that in general people think that proving the conditional true is as good a reason as proving it false, even though the former rationale is unsound. This may have something to do with the standard instruction to discover whether the statement is true or false, which suggests a symmetry between the two.

What do the 10% of high IQ solvers do differently? It cannot simply be that they understand falsification. They must also somehow overcome the powerful attentional bias towards the matching cards. We do know from Stanovich's (1999) work that such participants are much more able to suppress the influence of prior belief in reasoning, thus avoiding belief biases. Hence, as Stanovich argues, there is more to successful abstract reasoning than having effective analytic reasoning processes. It is important also to use volitional System 2 strategies to suppress actively or override the heuristic influences of System 1. This is well known by personal experience to psychologists who devise experiments on reasoning. In order to establish normatively correct responses we must consciously follow logical procedures that we have learned. If we relied on intuition, we might make the same errors as our participants.

## The thematic facilitation effect

The belief that thematic or realistic problem materials facilitated logical reasoning became widespread in psychology following the publication of the widely read book on reasoning by Wason and Johnson-Laird (1972). The first experiment that showed that realistic content could improve performance on the selection task was that of Wason and Shapiro (1971). The selection task used was on the face of it an indicative task in which people were asked to decide whether a claim was true or false. The claim was 'Every time I go to Manchester I travel by train'. Each card represented a journey with the destination on one side and the means of transport on the other. The visible sides were:

Manchester (p)   Leeds (not-p)   Train (q)   Car (not-q)

Note that the use of the quantifier *every* instead of *if* makes no difference to this task. An abstract control group given 'Every card which has a D on one side has a 3 on the other' produced 13% correct choices of p and not-q. The thematic group, however, chose the Manchester and car cards on 63% of occasions, a statistically significant difference.

The facilitation of selections on the 'towns and transport' problem initially appeared reliable, with some other published studies replicating the effect. Hence, the thematic facilitation effect for the indicative selection task became established. However, the phenomenon was cast into doubt when first Manktelow and Evans (1979) and then Griggs and Cox (1982) reported no facilitation at all with the same content (Americanized in the second study). It subsequently turned out that all realistic versions of the selection task that reliably facilitated performance also subtly altered the logic of the task, although the significance of this went unnoticed for many years in the psychological literature.

## The deontic selection task

An indicative conditional is an attempt to describe a matter of fact. In contrast, the aim of a deontic conditional is not description, but rather the expression of a rule to regulate or guide behaviour. Deontic conditionals express obligations or permissions and contain, either explicitly or implicitly, deontic modal terms, e.g. 'ought to', 'should', and the deontic 'must' or 'may'. Examples of deontic conditionals are the following:

If the traffic light shows red you must stop
If the traffic light shows green you may proceed
If you are under 18 years of age then you may not vote
If you are not a member of the club then you must not enter the building

Deontic inference is sometimes about whether such rules have been obeyed or violated. Experiments on deontic reasoning have generally investigated people's ability to make inferences about the violations of rules. Example of violations of the above are driving through red lights and entering a club building when not a member. Violations of deontic rules do not make the rules *false*. We will not get into a philosophical discussion about whether, or in what sense, such rules can be properly called 'true' or 'false'. But a deontic conditional can only be violated if it is properly in force as a rule: it has been laid down as a law or a regulation by a legitimate authority or someone with appropriate responsibility. Examples of legitimate authorities are the Ministry of Transport and a committee responsible for running the club. We are responsible for our own actions and can lay down moral or prudential rules for ourselves. For example, we can decide that we *should* go for a long walk the next morning if we have had too much to drink the night before. This would be a prudential rule for our own benefit. Just as we can sometimes violate a traffic law, we can violate one of our own deontic rules, even though we lay them down for our own benefit. We might also have the rule that we should not drink too much in the first place if we go to a party, but that does not mean that we never violate it.

We noted in Chapter 1 that deontic conditionals are closely related to decision making, and our account of the deontic selection task is decision theoretic, as we will explain below. Deontic conditionals are rules that authorities or responsible people lay down in the expectation of getting benefits, such as good heath, or avoiding costs, such as traffic accidents. An adequate account of deontic conditionals must, in our view, relate them to decision making about expected benefits and costs. But we begin by reviewing some important experimental results from deontic selection tasks.

## Experimental findings with the deontic selection task

The first published example of a deontic selection task seems to be the Postal Rule problem of Johnson-Laird *et al.* (1972). This was run on British participants who at the time were familiar with a rule that is no longer in force in Britain. Mail sent in sealed envelopes required a higher value stamp than unsealed mail, although the actual values were updated quite frequently in line with the high inflation rate of the time. The authors presented a problem congruent with this experience, but used Italian stamps with values in lira rather than pence. The rule was:

If a letter is sealed, then it must have a 50 lire stamp on it

A successful deontic selection task requires some minimal context, which was provided in this study by telling participants to imagine that they were post-office workers sorting letters. They were instructed to 'select those envelopes that you definitely need to turn over to find out whether or not they violate the rule'. This instruction provides another essential for a deontic selection task, the use of rule violation instructions.

Unlike subsequent replications, and unnoticed by most reviewers, the task presented *five* envelopes as shown in Figure 5.2. There were two envelopes with reverse side showing, one sealed (p) and one unsealed (not-p). There were three envelopes showing the address side, one with a 50 lire stamp (q), one with a 40 lire stamp (not-q), and one with no stamp, an extra not-q case. This blank envelope is the one usually ignored by later investigators. In this original study 88% of participants turned over the p and not-q envelopes compared with just 8% in the abstract control condition. The control task was an abstract indicative selection task.

While providing a dramatic result, this experiment was obviously open to criticism on the grounds of familiarity with a similar rule among the participant population (Manktelow and Evans, 1979). In fact, later research showed that such familiarity was a necessary requirement. Griggs and Cox (1982) found no facilitation using a Postal

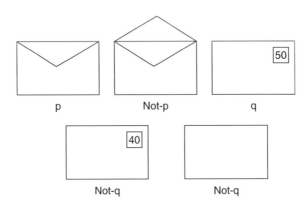

**Figure 5.2.** The five envelopes used by Johnson-Laird *et al.* (1972) in their Postal Rule problem.

Rule problem with an American population with no experience of such a rule. Cheng and Holyoak (1985) found facilitation for a Hong Kong group who had experienced such a rule, but not for an American control group who had not. Golding (1981) found more facilitation for older British participants who remembered the rule than for younger ones who had not experienced it. All of this suggests that, although the realistic content strongly affects responses on the task, it is not some process of formal reasoning that is being facilitated.

A similar objection can be made to another successfully facilitating deontic selection task, the Drinking Age problem of Griggs and Cox (1982). In this version people are asked to imagine they are police officers observing people drinking in a bar. The rule usually states that if people are drinking beer then they must be over 21 years of age (USA) or 18 years of age (UK). Most participants select under age people and beer drinkers for checking while ignoring those over age or drinking coke, but again this is fully compatible with their everyday experience. Deontic tasks have, however, been found to facilitate where people have no direct experience of the rule. One example is the Sears Problem described by Griggs (1983) in which people are asked to imagine that they are a manager in a department store and checking receipts for compliance with a rule such as, 'If the receipt is for over $50 then it must be signed on the back by the departmental manager'. Here people tend to turn over receipts that show over $30 (p) but not under (not-p) and those that are unsigned (not-q) but not those that are signed (q).

To cut a rather long story short, what happens in deontic selection tasks, with a rule 'if p then must q', is that most people will choose the p and not-q cards—those that could lead to detection of a violation—provided a number of conditions are met. The participants are familiar with the conditional, or its rationale as a rule for guiding behaviour is clear, or is clearly explained, to them. Facilitation with the Drinking Age Rule has also been shown to be ineffective if the minimal context is removed (Pollard and Evans, 1987) and to be weakened if the instruction to seek violators is removed in favour of an indicative true/false phrasing (Griggs, 1984). This last study (and other similar ones) also showed that using a search for violators instruction with unfamiliar problem content does not lead to facilitation.

Before turning to rival theoretical accounts of the deontic selection task, let us reflect on what the basic findings tell us about the nature of this kind of hypothetical thinking. The requirement of familiarity and minimal context suggests the operation of subtle and powerful pragmatic influences on people's imagination. If the context is such that people can appreciate the significance of violating cases, then they have no difficulty in picking the correct cards. A variety of experimental procedures and content manipulations have been introduced in selection task experiments to show that foregrounding counter-examples to the conditional will facilitate card selections (Pollard and Evans, 1981; Pollard and Evans, 1983; Green, 1995; Love and Kessler, 1995).

In discussing the abstract indicative selection task, we came to the view that participants do engage in conscious hypothetical reasoning, although most fail to solve the task because they overlook cards that are not heuristically cued and they fail to understand the distinction between verification and falsification. The fact that the minority of solvers have high general intelligence scores (Stanovich and West, 1998a) also suggests that analytic reasoning is required for success on the indicative task. Is the same true on the deontic task? It seems not. Stanovich and West also showed that there were much smaller differences in intelligence scores between solvers and non-solvers of the deontic selection task. From a dual process theory perspective this makes sense, as everything points to the operation of rapid pragmatic processes that make the counter-example case available without the need for conscious analytic reasoning. To be sure, a little reasoning is then needed to translate this into the correct card selections in the context of the instructions, but this is not hard enough to give any great advantage to high participants with very high general intelligence.

## Pragmatic reasoning schemas

The earliest type of theory proposed for the deontic selection task is one implicit in the above discussion, which was termed the 'memory-cue' hypothesis by Griggs and Cox (1982) and more generally 'availability' theories by Cosmides (1989; see also, Pollard, 1982). The idea is that pragmatic or associative processes related to the problem content or context cue the retrieval from memory of the counter-example case, thus facilitating choice of the correct cards. Closely related are accounts in terms of relevance (Evans, 1995; Evans and Clibbens, 1995) or in terms of mental model theory where people are supposed to focus on the explicit content of mental models (Johnson-Laird and Byrne, 1992; Johnson-Laird, 1995).

We have already seen that these types of theory are quite well supported by the basic evidence on the deontic task. However, deontic rules facilitate in some contexts where people do not have experience of the problem content. In these cases, people have experience of similar or analogous contexts, or the rationale of the rule is explained to them. Hence, some more sophisticated pragmatic mechanism is required than simple associative cueing.

In response to this problem, Cheng and Holyoak (1985) proposed a theory of *pragmatic reasoning schemas*. The idea is that people have acquired, through experience, sets of rules for reasoning about particular kinds of situations, clustered together in schema. The schema is abstract but domain-sensitive and hence to not be confused with rule-based mental logics of the kind discussed in Chapter 4. Cheng and Holyoak's permission schema had the following four production rules:

Pr1—If the action is to be taken, then the precondition must be satisfied
Pr2—If the action is not to be taken, then the precondition need not be satisfied
Pr3—If the precondition is satisfied, then the action may not be taken
Pr4—If the precondition is not satisfied, then the action must not be taken

A schema like this will facilitate reasoning *provided* that the schema is retrieved from memory, has its variables instantiated with the current content and is applied in reasoning about the problem. Consider some of the findings discussed earlier. The Drinking Age Rule of Griggs and Cox (1982) is normally a reliable facilitator. This is a permission rule where the action is drinking beer and the precondition is being over the legal drinking age. However, Pollard and Evans (1987) showed that presenting the rule without the minimal police officer scenario removed the facilitation. Griggs and colleagues have shown in several experiments that removing the violation instruction greatly weakens the effect (see Evans *et al.*, 1993, pp. 104–7). These could be interpreted as factors that would affect the likelihood of the permission schema being retrieved and applied to the problem.

The schema theory implies that people reason about each card in turn, applying the relevant production rules. With the permission schema, only rules Pr1 and Pr4 trigger actions (choosing p and not-q cards) as they contain the *must* term. However, the theory would seem to imply that participants think about each card. As Evans and Clibbens (1995) pointed out, this is inconsistent with the findings applying the card inspection time method to facilitatory permission rules (Evans, 1996). A good deal of controversy also resulted from the claim of Cheng and Holyoak (1985) that an *abstract* permission rule, 'If one is to take action A then one must first satisfy precondition P', facilitated correct choices. The problem was the introduction of a potential confounding factor— the use of explicit negatives on the card, a factor later shown by Evans *et al.* (1996a) to remove matching bias on the abstract selection task. Later experiments provided clear evidence that the facilitation was indeed dependent upon the use of such explicit negatives (Jackson and Griggs, 1990; Noveck and O'Brien, 1996).

The main problem with Cheng and Holyoak's schemas for us in this book is that they tried to explain deontic conditionals in terms of other conditionals, namely, the above production rules, Pr1–Pr4. They did not specify a full logical system for their production rules nor for the deontic modals, 'must' and 'may', that these rules contain. Their account could be called circular because they make use of the deontic modals to try to explain reasoning with deontic conditionals. But the theoretical questions we have about ordinary deontic conditionals, and ordinary deontic modals, become theoretical questions we have about Cheng and Holyoak's production rules as conditionals that contain deontic modals. We wish to gain a deeper understanding of conditional thought, and it does not get us very far to replace one conditional with another. Moreover, Cheng and Holyoak's schemas are simply designed to account for responses in the selection task. They do not explain why deontic conditionals are asserted in the first place (Over *et al.*, 2004). We seek this explanation, of course, in a theory of hypothetical thought.

Consider the general type of postal rule we discussed above, 'If a letter is sealed, then it must have a high value stamp on it.' Why would a postal authority adopt such a rule in the first place? Why might the authority later withdraw the rule? These questions are not answered, or even addressed, by Cheng and Holyoak's schema theory. For us, these are questions about hypothetical thought. Our approach is to modify what Ramsey (1931/1990, p. 247) suggested about indicative conditionals (see Chapter 2). Recall that Ramsey said that people who are arguing about whether 'if p then q' holds do this by '. . . adding p hypothetically to their stock of knowledge and arguing on that basis about q . . .' Consider how authorities at the Post Office would try to determine, say, in a committee meeting, whether to lay down the Postal Rule or to withdraw it. Our view is they would hypothetically suppose that a letter is sealed, and then argue about the expected benefits, or the expected costs, to the Post Office of charging more to deliver it. We will elaborate on this view in the section below on decision theoretic explanations of the deontic selection task. We hold that this type of explanation in hypothetical thought is an advance beyond trying to explain ordinary deontic conditionals in terms of production rule conditionals. Cheng and Holyoak's schemas might find a place in a theory of hypothetical deontic thought, but it would be a derived place and not a primary one.

## Darwinian algorithms

Cosmides (1989) introduced evolutionary psychology into the reasoning field with the novel claim that facilitation on the deontic selection task resulted from Darwinian algorithms. These algorithms are supposedly highly content-specific mechanisms in innate mental modules. The original algorithm proposed was one for detecting 'cheaters' on 'social contracts', i.e. people who got benefits without paying an agreed cost by violating social agreements. Cosmides (1989) argued that it had been adaptive for our evolutionary ancestors to have had an increasing ability to identify cheaters in this sense. She also claimed that the only efficient way for this ability to exist would be as a dedicated, content-specific innate module, the sole function of which was to identify cheaters. She implied that facilitation was found only in deontic selection tasks where there was the possibility of detecting cheaters on social contracts. When Manktelow and Over (1990) later showed facilitation with prudential rules where violators are not cheaters, e.g. the hospital rule, 'If you clean up spilt blood then you must wear rubber gloves', Cosmides' response was to propose a second innate algorithm for hazard management (Cosmides and Tooby, 1992; Fiddick *et al.*, 2000).

We have criticized Cosmides' theory before (Manktelow and Over, 1991; Evans and Over, 1996a; Over, 2003) and will not repeat our points in detail here. It is most implausible that a high level ability like using deontic conditionals could be fully explained in terms of innate domain-specific modules. Many of these conditionals express social, legal, and moral rules that could not have greater cultural richness. (See Holyoak and Cheng, 1995, for example, on reciprocal rights and duties.) The sophisticated rules that lie at the basis of scientific research itself are deontic, such as how we *should* design experiments, and what we *may*, and *may not*, infer from significant results. The hypothesis of specific innate mechanisms for understanding deontic rules is a very strong claim that can be accepted only if more parsimonious accounts are eliminated first. And we certainly cannot eliminate the possibility that a general-purpose learning mechanism has evolved by natural selection and helps to account for deontic thought (Almor and Sloman, 1996; Almor, 2003). Of course, schema theorists were quick to point out that their theory, depending upon no evolutionary argument, could account for the same findings (Cheng and Holyoak, 1989).

The other major problem for Cosmides lies in the nature of her experiments. The experiments reported by Cosmides (1989) made use of very long and complex verbal scenarios that differed in numerous ways from the control conditions with which they were compared, rendering interpretation of causal factors almost impossible. Although the more recent experiments of Fiddick *et al.* (2000) are better controlled, they have been subjected to a strong critique by Sperber and Girotto (2002), who argue that both these experiments and the earlier ones by Cosmides (1989) contained instructions that changed the nature of the task to one that is trivially simple. This change did not apply to the standard selection tasks with which they were compared. Sperber and Girotto argue that because the instructions included a specific request to detect cheaters, the task becomes a simple classification task, without the inferential demands of the normal selection task.

Consider one of Cosmides (1989) original problems that used the rule (in an elaborated context) where you are promised by Big Kuku, 'If you get a tattoo on your face, then I'll give you a cassava root.' Although participants were asked to look for violations of this rule, they were also explicitly instructed to check whether Big Kuku was a cheater. Identifying the cheating case—tattoo with not cassava root—is obviously simple. Are the card selections then just classifying the cheating cases? Sperber and Girotto show in their experiments that simply asking people to identify cheaters, whether or not accompanied by instructions to search for violations, leads to p and not-q choices. However, there is nothing special about cheaters. The same kind of instructions facilitates choices for any concept that is being classified because it changes the nature of the task. For example, in one experiment participants were told that Paolo collected pictures of gliders that were defined as moving in the air without an engine. The four cards were:

Moves in the air   Moves on rails   Has engine   Does not have engine

Participants were instructed to decide which cards to turn over in order to see if there were any gliders and of course they chose the first and last cards. However, their point was that the cheater detection task trivializes the selection task in the same way. Once you have indicated by the context that, say, a cheating case is one with a tattoo and without cassava root, choosing from the following four cards is just as simple:

Tattoo   No tattoo   Cassava root   No cassava root

The inferential difficulty of the identifying the not-q card in the normal selection task is thus removed by the instruction to detect cheating cases. Sperber and Girotto present several well-designed experiments to support their argument.

On the positive side, Cosmides and her collaborators do relate deontic conditionals to benefits and costs. It is clear that there are benefits, for reproductive success and more widely, in avoiding the costs of being cheated and of being exposed to possible hazards. But the most negative aspect of Cosmides' whole approach is that in it the word 'if' is not the trigger for hypothetical or any other sort of thought. The occurrence of this word apparently does nothing in general. It is the content of the words that come after 'if' that cause a cheater detection, hazard identification, or whatever content-specific module to come on line and do the work. This approach has no general theory of 'if', not even of its general use in deontic discourse, and we consider that unacceptable (but see also Cosmides and Tooby, 2000; Over, 2003).

## The relevance theory of Sperber *et al.*

A more promising account of the deontic selection task, in our opinion, is that presented by Sperber *et al.* (1995). It is worth our while to give this some attention as it makes some detailed proposals about the communicative function of conditionals. This was developed from the well-known relevance theory of Sperber and Wilson (1995) applied to explain pragmatics in language and communication. In its most recent version the theory has two principles of relevance.

1. The first (cognitive) principle of relevance:
   - the greater the cognitive *effect* resulting from processing the information, the greater its relevance;
   - the greater the processing *effort* required for processing information, the less its relevance.
2. The second (communicative) principle of relevance:
   - every utterance conveys a presumption of its own relevance.

Those familiar with the original version of the theory (Sperber and Wilson, 1986) will recognize the communicative principle as the one then described as *the* principle of relevance. The communicative principle, which some authors prefer to describe in terms of Gricean maxims (Grice, 1975) is certainly useful in understanding research on human reasoning, especially with thematically rich content, as we shall see later in the book. The cognitive principle, with its notion of processing effort is, however, central to the explanation of conditional reasoning given in this paper.

The authors proposed that a conditional may be processed to one of three levels of increasing depth (and effort) which they call cases (a), (b), and (c). Level (a) conveys nothing more than Modus Ponens (there is always a q with a p). At level (b), the conditional is read as asserting that there are some p type cases and therefore some q type cases, i.e. that cases of pq exist. Logically, the conditional does not assert that there are cases of p, but it would be a violation of the communicative principle of relevance to assert a conditional where no cases exist to which it could be applied, so with only a little more processing effort this pragmatic inference can be derived. Level (c) requires the most processing effort, explicitly representing the conditional as ruling out cases of p and not-q. However, we do not think an instance of an indicative conditional, 'if p then q', completely rules out possible TF cases. It is rather that the possible TF cases affect the conditional probability of q given p, $P(q|p)$, and so the probability of 'if p then q' (see Evans *et al.*, 2003a).

Sperber *et al.* (1995) argue that abstract indicative versions of the selection task, with their minimal context, do not generally induce representations deeper than level (b). Hence, the assertion, 'If a card has an A on one side then it has a 3 on the other', orients people to think about cards with A instances and 3 instances, leading to the usual matching response. However, a deontic context, especially with the rule violation instructional set, can induce a deeper reading of the conditional in which cases of p and not-q are represented as violations. This happens when the context provides a specific cue to the relevance of such a counter-example case, such as an instruction to identify cheaters. Once such a representation is formed, the problem becomes simple in the same way (Sperber and Girotto, 2002). For example, given the Drinking Age Rule, 'If a person is drinking beer then that person is over 18 years of age', together with a police officer scenario, the participant will build the representation that an underage drinker is a violation, after which the p and not-q cards select themselves.

What is particularly interesting about the Sperber *et al.* (1995) account (supported by a set of interesting experiments that we do not have space to describe) is that it gives a *domain general* account of conditional inference in contrast with domain-specific theory, such as pragmatic reasoning schemas and Darwinian algorithms (see Fiddick *et al.*, 2000, for a lengthy attempt to refute these proposals). Content and context influence representations by the principles of relevance that are quite general. Depending on the context, realistic conditionals may or may not be read to level (c). For example, if an employer says to an employee, 'If you stay an extra hour tonight, I will pay you double time', this would probably processed at level (a), simply inducing the listener to think about the money under the hypothesized condition. They were not expected or required to work the extra hour anyway, so there is nothing in the context to induce the notion that the conditional forbids anything. On the other hand, if an employer says, 'If you want to keep this job, you will wear a tie to work every day', this would certainly be read as level (c), that is as an assertion that you cannot keep the job and not wear a tie. Part of the context that would contribute to processing this sentence would (presumably) be the fact that the employee had been turning up to work without a tie, and had probably been told off for it before, and so on.

Sperber *et al.*'s (1995) account is consistent, of course, with the evidence we reviewed earlier in which the availability of counter-example p and not-q cases and orientation towards detecting them seemed to be key factors in facilitatory versions of the deontic selection task. It is more parsimonious than either the pragmatic reasoning schemas or Darwinian algorithm theories. It does not require domain-specific reasoning apparatus and of course it does not need to posit domain-specific innate modules either. Sperber *et al.* recognize that there should be a general theory of 'if', and also that relevance can be affected by benefits and costs. However, we believe that deontic thought must be seen as an instance of hypothetical thought and related much more deeply to expected benefits and costs.

## Decision theoretical accounts

A number of authors, including us, have proposed that participants approach the selection task as a decision-making problem and so are influenced by the utilities and probabilities of the possible outcomes (Manktelow and Over, 1991, 1995; Over and Manktelow, 1993; Oaksford and Chater, 1994, 1995; Kirby, 1994; Nickerson, 1996; Evans and Over, 1996a). Work in this tradition that makes use of the indicative selection task, and attempts to influence probabilities by experimental procedures, will be discussed in Chapter 8 (on probabilistic treatments of conditionals). We briefly consider here the influence that perceived goals and subjective utilities has in the deontic selection task. The decision context and perceived goals are crucial pragmatic factors that influence the way the conditional is represented and the task is approached.

Like Sperber and colleagues, we have long been interested (since Evans, 1989) in the idea of relevance in reasoning tasks. For indicative conditional reasoning and the indicative selection task, we believe that relevance can be equated with what we term *epistemic utility*: the subjective value to people of the knowledge and information that they wish to gain (Evans and Over, 1996a,b). The use of deontic expressions in a scientific context, as when we say how an experiment should be designed, can be about epistemic utility. But in most deontic reasoning and tasks, relevance comes from benefits and costs in the ordinary sense. The benefits to be gained, or costs to be avoided, concern such goals as good health, e.g. when underage alcohol drinking is forbidden, or making a profit in business, e.g. when signed receipts are required to prevent counterfeiting. We adopt a deontic conditional of the form, 'if p then must q', in the first place because of a judgement about benefits to be gained or costs to be avoided. Supposing p in hypothetical thought, in a kind of deontic Ramsey test, we would judge there to be a higher expected benefit, or a lower expected cost, from q than from not-q. For instance, supposing that we are cleaning up spilt blood in a hospital context, we foresee an expected (hygienic) benefit from wearing rubber gloves and an expected (life-threatening) cost in not wearing them. We therefore decide to have the rule that, if we are cleaning up spilt blood, then we must wear rubber gloves. Such decision theoretic reflection in hypothetical thought is the first way our account of deontic conditionals and modals is decision theoretic (Over *et al.*, 2004).

The second way our account is decision theoretic relates more directly to the selection task. The familiar context of a deontic selection task, or its rationale, can suggest benefits to be gained or costs to be avoid by finding violations of the given deontic conditional, which is presupposed to have been already adopted by some relevant authority. The participants in deontic tasks have to decide which cards to turn over to reveal violations of the deontic conditional. The familiar context of the task, or its rationale, makes it clear to the participants that they will gain a benefit from finding, or suffer a cost from failing to find, violations. It is striking how often experiments on deontic selection tasks ask the participants to imagine that they are doing a job, e.g. as postal workers, police officers, or store managers. In the context of the task, the participants are not doing their imagined job properly when they fail to find violations of the given deontic conditional. Most people have, direct or indirect, experience of the disciplinary action that can follow the failure to do a job properly. Of course, deontic selection tasks can be about other possible costs, e.g. to good health. There may be a tendency for people to have some loss aversion in deontic tasks and to focus on possible costs more than possible benefits (Manktelow and Over, 1990). However, little real reasoning, in System 2, is needed for people to identify possible benefits and costs in deontic selection tasks where facilitation occurs.

## Conclusions

Research on the selection task is very interesting with regard to pragmatics of 'if', although it is a poor vehicle for its traditional purpose of investigating deductive reasoning. On abstract and indicative versions of the task, we have shown that analytic (System 2) reasoning does occur, but does not normally exert much influence on the choices made, although it affects the amount of time spent on the decision. Heuristic processes (based in System 1) focus attention in such a way that choices greatly underestimate people's actual deductive competence. In particular, the not-q card is typically chosen by only about 10% of university students (those of unusually high general intelligence), whereas the equivalent Modus Tollens inference on the conditional inference task (Chapter 3) is made at least 70% of the time in most studies. There is a big difference between drawing a correct inference from some presented information and determining independently that the information is relevant in the first place.

The deontic selection task, on which most studies have focused in the past 20 years or so, provides little evidence of deductive reasoning and shows little advantage for high IQ participants. What research on this task shows is that hypothetical thinking about conditional statements that are both deontic and embedded in context is facilitated by rapid and automatic pragmatic processes to do with expected benefits or costs. The recent claim of Sperber and Girotto (2002) that asking people to detect cheaters trivializes the task can be extended by analogy to most contexts for deontic selection tasks. The reasoning is simple because the context conveys both the counter-example case and its relevance, in terms of possible benefits and costs, to the task set. The standard presentation of the drinking age problem, for example, has no reference to cheaters, but is only fully effective if (a) the minimal police officer scenario is present, and (b) the instructions refer to detecting violations of the rule. This presentation is sufficient to trigger the identification of underage alcohol drinkers as the relevant case to be uncovered to do the police office's job properly.

The lesson to be learned from this literature, perhaps, is that the ordinary processing of conditional statements calls on far less analytic, System 2 reasoning than psychologists and philosophers have traditionally assumed. In the absence of helpful context—as in the typical abstract indicative task—hypothetical thought focuses attention rapidly on the antecedent and the expectation of p and q confirming cases. There is a susceptibility to matching bias that usually disappears when context is present, *even when that context does not facilitate correct choices* (see Evans, 1995). People of high intelligence can more easily make use of System 2 to solve indicative tasks correctly. People tend to be more equal in their ability to solve deontic selection tasks thanks to powerful pragmatic and System 1 processes that focus attention on cases in which there are expected benefits to be gained or costs to be avoided.

In the next chapter, we consider a range of psychological studies that have investigated contextual effects on thematic conditionals on problems other than the selection task. These tasks have predominantly followed the deductive reasoning paradigm, by asking people to assume premises and draw necessary conclusions. Given our analysis of the selection task, however, it will be interesting to see whether these tasks are approached deductively or dominated also by pragmatic processes. We shall return to say more about our accounts of indicative and deontic conditionals in Chapter 9.

# 6  Conditionals in context

Ordinary conditionals are not asserted in isolation. They are used in a context where a speaker communicates with a listener. This context includes the goals and motivations of the speaker and via his/her theory of mind those of the listener and the shared knowledge that they possess. We have several times already talked of the importance of pragmatics in understanding the use of conditionals and have looked at some experimental work on how context influences reasoning or judgement in the previous chapter on the Wason selection task. In this chapter, we focus mostly on content and context effects on the conditional inferences Modus Ponens (MP), Denial of the Antecedent (DA), Affirmation of the Consequent (AC), and Modus Tollens (MT) that were discussed with reference to experiments on arbitrary materials in Chapter 3.

We have already discussed some of Sperber and Wilson's (1995) work on relevance theory to account for the pragmatics of conversation and communication. In particular, we have considered the communicative principle of relevance, namely that all communications include a guarantee of their own relevance. Some authors (for example, Hilton, 1995) prefer to apply the more elaborated earlier scheme of Grice (1975), who set a series of maxims or rule of conversation based upon the assumption of co-operation between speaker and listener. These include the requirements to be co-operative, relevant, and truthful. Of most interest for our purposes, however, is the maxim of quantity:

- Make your contributions as informative as required for the current purposes of the exchange.
- Do not make your contribution more informative than is required.

We can paraphrase the maxim of quantity as follows: say as much as you mean, but say no more than you mean. Grice also used the term 'implicatures' to convey implicit or pragmatic inferences, in contrast with 'explicatures', which are the logical implications of utterances. In the psychological literature, the former are often referred to as 'invited inferences'. (One of the earliest authors to apply Gricean principles to the issue of invited inferences with conditionals was Fillenbaum, 1975, 1976.) Hence, the common tendency of participants in psychological experiments to endorse the conditional fallacies, DA and AC, may be accounted for by mental logicians on the grounds that they are invited inferences induced by the context (Braine and O'Brien, 1998a).

Consider the application of the maxim of quantity to the form, 'if p then q'. The consequent q is asserted conditionally upon p. According to the maxim, it should not have been asserted conditionally if it could have been asserted *unconditionally*, as otherwise you would be saying less than you mean. So in many real world contexts, while 'if p then q' carries the logical implication that p is sufficient for q, it will also carry the implicature that p is necessary for q. Consider the following example of a conditional promise, issued by mother to son:

6.1   If you tidy your room then I will take you to the cinema

The inverse (or DA) inference is strongly invited here: if you do not tidy your room then I will not take you to the cinema. Why make the promise conditional if you intend fulfilling it in any event? Note also that this inference seems more strongly invited than the AC inference 'I take you to the cinema therefore you tidied your room' even though DA and AC would be equally entailed by biconditional truth table. We believe this is because people generally reason forwards in time. Conditionals of this kind are discussed in detail later in this chapter.

In spite of the above analysis, we should appreciate that necessity is not always pragmatically suggested by thematic conditionals. What is relevant depends always on the perceived goals of the listener. Sometimes we are only interested in sufficiency: we desire q and need to find an action p that will produce it. Thus the advice 'If you take the train then you will get to your appointment on time' is relevant to the extent that it provides a solution to a problem. It need not be the only solution. Note that this advice is not an inducement: the speaker is not trying to change your behaviour, but merely intends to be helpful.

We should also point out that pragmatic inference is *defeasible* (Oaksford and Chater, 1991; Elio and Pelletier, 1997). That is to say, pragmatic inferences drawn from some given information may be withdrawn in the light of later information. It is understandable perhaps that implicatures or pragmatic inferences should be updated. Consider, the conditional promise

6.2   If you pass the exam then I will buy you a new bicycle

Suppose the child to whom this promise is given discovers a brand new bike hidden in the garden shed before sitting the examination. She might well decide that this is intended for her and cancel the previously made DA inference. However, as we shall see later in this chapter it is not only implicatures but also explicatures that can be cancelled by new information, suggesting that this distinction is weaker than it appears.

Much of the psychological work on conditional inference in context has used the deduction paradigm in which people are instructed to assume the premises and to draw conclusions that necessarily follow. It is quite doubtful as to whether this is appropriate (see Evans, 2002a) as it necessarily defines pragmatic influences as sources of bias and error. Some studies, however, have used pragmatic inference instructions where people are simply asked to decide what they think follows and are even in some cases allowed to express their conclusions with a degree of confidence or probability. It turns out the form of instructions used has quite a major influence on the reasoning observed in some of these studies. In reviewing the evidence below, we will pay careful attention to the type of instructions used.

We should also note that many of the psychological experiments on 'thematic' conditionals considered later in this chapter do not supply an actual context for them within the experiment (although some do). However, the mere use of conditionals with realistic terms that have an identifiable connection is sufficient for pragmatic factors to operate: the context is simply supplied by the participants from their store of knowledge and belief. Thus people have no difficulty in identifying a statement such as 'if the key is turned then the motor will start' as causal, or the statement 'if you drink beer then you must be over 18 years of age' as a permission rule and so on, and they can equally easily retrieve and apply relevant knowledge to such conditionals.

Before examining psychological studies of conditional inference in context, we will briefly discuss a phenomenon known as *belief bias*. Most of the studies in this literature have used categorical syllogisms rather than conditional inferences. However, it is important to understand the belief bias phenomenon as it certainly will affect conditional inferences also and must be taken into account in the interpretation of some of the studies that we later review.

## The belief bias effect

The so-called belief bias effect in deductive reasoning was first reported by Wilkins (1928). The idea was that the prior beliefs bias people's assessment of the validity of logical arguments. Instead of deciding whether a conclusion follows logically from some given information, people might instead judge an argument on whether they believe its conclusion. There were several reported replications of Wilkins's findings but these early studies were affected by a host of methodological problems (Evans, 1982). However, the nature of the phenomenon was clearly established, with relevant controls, by Evans *et al.* (1983). In this study, participants were given syllogisms to evaluate, comprised of two premises and a conclusion and asked to say whether the conclusion followed. They were given clear deductive reasoning instructions: 'You should answer the question on the assumption that the information given . . . is, in fact, true. If the conclusion necessarily follows from the statements . . . you should answer "yes", otherwise "no".'

Evans *et al.* devised syllogisms that fell into four categories, according to whether the conclusion was valid or invalid, and whether it was believable or unbelievable. In this study, invalid conclusions, or fallacies, had conclusions that *could* be true given the premises, but did not have to be. Examples of the four types are shown in Table 6.1, together with the frequency (over three experiments) with which participants said that the conclusion necessarily followed. Three statistically reliable findings emerged that have been replicated in most subsequent studies: (1) people endorse more valid and than invalid arguments; (2) people endorse more believable than unbelievable conclusions; and (3) the effect of belief is stronger for invalid than valid arguments.

Thus it can be seen that university students, clearly instructed to assume the premises and draw conclusions that necessarily follow, nevertheless show a substantial influence of belief. The interaction—finding (3)—is also important. The belief effect is stronger on fallacies. Evans *et al.* produced evidence that belief and logic were set in conflict with individual participants, effectively competing for control of their responding, a finding in line with the dual process theory of reasoning (see Chapter 1). Subsequent research has shown that the ability to resolve belief-logic conflict problems (valid-unbelievable and invalid-believable) in favour of logic is related to high general intelligence (Stanovich and West, 1997) and declines sharply in old age (Gilinsky and

**Table 6.1.** Evidence of belief bias in reasoning, from Evans *et al.* (1983). Example syllogisms with % acceptance of conclusions as valid

| | |
|---|---|
| Valid-Believable | |
| No police dog are vicious | |
| Some highly trained dogs are vicious | |
| Therefore, some highly trained dogs are not police dogs | 89% |
| Valid-unbelievable | |
| No nutritional things are inexpensive | |
| Some vitamin tablets are inexpensive | |
| Therefore, some vitamin tables are not nutritional | 56% |
| Invalid-Believable | |
| No addictive things are inexpensive | |
| Some cigarettes are inexpensive | |
| Therefore, some addictive things are not cigarettes | 71% |
| Invalid-unbelievable | |
| No millionaires are hard workers | |
| Some rich people are hard workers | |
| Therefore, some millionaires are not rich people | 10% |

Judd, 1994). There is even neuropsychological evidence that resolution in favour of logic or alternatively belief is differentiated in terms of the locus of brain activity (Goel and Dolan, 2003).

The description of the belief effect as a *bias*, presupposes that people are endorsing fallacies because they believe the conclusion. In fact, people have a very high tendency to endorse such fallacies with abstract problem materials (Evans *et al.*, 1999a). When belief bias experiments include problems with neutral conclusions, then it is shown that the effect of belief is really a tendency to suppress fallacies with unbelievable conclusions, rather than facilitate them with believable conclusions (Newstead *et al.*, 1992; Klauer *et al.*, 2000; Evans *et al.*, 2001). This tendency could be regarded as a debias! However, the (smaller) belief effect on valid arguments, causing reduced acceptance with unbelievable conclusions, certainly is a bias relative to the instructions given. Although weaker than the effect on fallacies, this effect was observed by Evans *et al.* (1983) and in most replication studies. Evans *et al.* (1994) attempted to remove the belief effect by strong instructional emphasis on logical necessity and succeeded in reducing, but by no means eliminating, the influence of belief by this manipulation. Relaxing deduction instructions can greatly increase the influence of belief and pragmatic factors, as we shall see in reviewing conditional inferences studies below.

One factor not controlled by Evans *et al.* (1983) was the believability of the premises. Perhaps people confuse validity with *soundness*. A *sound* argument can be defined as a valid argument with true premises. Direct evidence that people prefer sound arguments when asked to judge validity in syllogistic reasoning has been shown by Thompson (2001). Some experimenters have tried to control for premise believability by using nonsense linking terms in the premises, so that each premise is belief neutral (Newstead *et al.*, 1992) as in:

All sparrows are haemopheds
All haemopheds are birds
Therefore, all sparrows are birds

With such controls, belief bias effects are still observed. However, in spite of this, there is still a strong argument that belief bias may reflect a preference for sound arguments (Evans *et al.*, 2001). The difficulty is this: a sound argument can never, by the definition of validity, have a false conclusion: the conclusion of a valid inference must be true given that the premises are true. Hence, if a conclusion is unbelievable, we may reject a valid argument immediately as unsound without looking closely at the premises. We can simply decide, a priori, that this argument must be based on false premises, even if that falsity is only implicit. This issue of soundness turns out to be very important to the interpretation of some of the context manipulations with conditional inferences discussed later in this chapter.

Research on belief bias in syllogistic reasoning is very interesting and important. It shows us that notwithstanding the presence of clear deductive reasoning instructions, people are strongly influenced by the problem content and their attitude towards it. People are clearly making an effort at deduction in accord with the instructions, as they prefer valid conclusions overall and reason more logically than when such instructions are present. Nevertheless, the pragmatic influence of prior belief is very powerful and hard to suppress and only those with very high general intelligence seem able consistently to reason in accordance with abstract principles.

## Necessity and sufficiency

The 'chameleon' theory of conditionals (Braine, 1978) recurs regularly in the literature on conditional reasoning, although it is rarely given this name. The idea is that conditionals may be read as conditional or biconditional according to context. This means that we should expect some participants to make the inferences valid only for conditionals, MP and MT, and some to make all four: MP, DA, AC, and MT according to which reading they adopt. From the evidence reviewed in Chapter 3, we already know that this theory cannot be used to define the interpretation of abstract indicative conditionals as truth functional. It is true that the fallacies DA and AC are commonly endorsed, but we showed there that the pattern of inference rates is not consistent with either truth table. We also showed that inferences with abstract conditionals were subject to systematic biases, such as matching bias and double negation bias. It is not surprising, then, that early psychological studies seeking truth functional patterns among individual reasoners on such tasks were largely unsuccessful (see Evans, 1982, pp. 135–137).

A subtler version of the chameleon theory is that people adopt a conditional or biconditional interpretation, but that the observed reasoning is not truth functional due to psychological factors affecting the difficulty of reasoning. For example, we have already shown in Chapter 4 that both mental logic and mental model theorists proposes mechanisms of reasoning that yield MT as a valid inference, but supplement these with psychological accounts that explain why MT in practice is drawn less frequently than MP. The biconditional reading that some participants may adopt would add the conditional 'if q then p' to the representation of 'if p then q'. However, this version would have to predict that the ratio of MP to MT would be the same as the ratio of AC to DA, if fallacies results from analogous reasoning, and this does not fit the data with abstract indicative conditionals either. Nevertheless, the results do fit some rough notion of the theory in that the valid inferences are usually highly endorsed in all studies, and the fallacies are quite variable across studies, suggesting that some experimental procedures encourage biconditional readings more than others.

The underlying issue here concerns sufficiency and necessity relations. The conditional specifies the following links, which entail the valid inferences:

p is sufficient for q (given p as a minor premise, you must have q, MP)
q is necessary for p (given not-q as a minor premise, you must have not-p MT)

The biconditional also asserts the following links:

q is sufficient for p (given q as a minor premise, you must have p, AC)
p is necessary for q (given not-p as a minor premise, you must have not-q, DA)

In Chapter 3, we provided evidence for defective truth tables. The defective conditional assigns TT as true, TF as false, but FT and FF as irrelevant. The defective biconditional assigns TT as true, TF as false, FT as false, and FF as irrelevant. We showed there that MP and MT are valid for the defective conditional (it is enough to know that you cannot have the state of affairs, TF). Similarly, the inferences DA and AC are valid also for the defective biconditional, following from the fact that you cannot have FT. So any evidence for conditional and biconditional patterns of conditional inference would be inconclusive with regard to the issue of whether underlying truth tables were material or defective. It is important to bear this in mind.

Now what of thematic conditionals presented with realistic content or in context? These are not necessarily subject to the same cognitive biases as abstract conditionals, and indeed there is good evidence that matching bias does not operate with realistic conditionals (see Evans, 1998b, for a review of relevant studies). So perhaps the chameleon theory will fare better with such conditionals. A relevant study is that of Hilton *et al.* (1990) who examined conditional inferences with causal conditionals and attempted to manipulate the perception of sufficiency and necessity relations by providing additional information. For example, with the sentence:

6.3   If the exam is easy then he will pass

the words added were (he works hard) for some participants and (he does not work hard) for others. The former provides an alternative cause for the consequent, generally known as an *alternative antecedents* manipulation. Hilton *et al.* assume that the default reading of a causal conditional will be biconditional: p is both sufficient and necessary for q. They hoped to shift people towards a conditional reading in conditions where an alternative antecedent was suggested. In the above case it may not be necessary for the exam to be easy in order for him to pass, as he is working hard.

Hilton *et al.* found, in line with the earlier studies reviewed by Evans (1982) that most inference patterns (58%) produced by individual participants did not fit consistently with either a conditional or biconditional truth table. When patterns were classified by the nearest truth table, however, the predicted trends were observed. People treated causal statements as less biconditional when alternative antecedents were specified, a result actually first demonstrated by Markovits (1984). It is interesting to note that Hilton *et al.* used instructions somewhere between pragmatic and deductive. Participants were told to treat the conditional as if it had been uttered in conversation and that the second piece of information (minor premise) was something they could treat as being known to the case. They were given a conclusion to evaluate as true, false, or sometimes true and sometimes false. They were not explicitly told to judge necessity. It is also worth noting that when additional information is provided by the experimenter, as in their parenthetical text, the conversational implicature is that this is relevant to the task and should be taken into account.

Alternative antecedents undermine the necessity of p for q. It is also possible pragmatically to undermine the sufficiency of p for q by introduction of *disabling conditions*. Disabling conditions suggest reasons why p may not lead to q. One method of approaching this issue is to *correlate* people's perception of necessity and sufficiency relations on particular sentences with their conditional inference patterns for the same sentences. Cummins *et al.* (1991) reported an exceptionally well designed study on causal conditionals that it is worth considering in a little detail. First, they conducted a pre-test with independent participants from the main reasoning experiment. The purpose of this was to establish the ease with which conditionals in causal contexts could bring to mind either disabling conditions or alternative antecedents. This divided into two tasks which we illustrate with examples.

*Alternative antecedents generation*
*Rule*: If Joyce eats candy often, then she will have cavities
*Fact*: Joyce has cavities, but she does *not* eat candy often
Please write down as many circumstances as you can that could make this situation possible

*Disabling conditions generation*
*Rule*: If Joyce eats candy often, then she will have cavities
*Fact*: Joyce eats candy often, but she does *not* have cavities
Please write down as many circumstances as you can that could make this situation possible

By administering these tasks for a large number of conditionals, the experimenters were able to select thematic conditionals falling into four categories classified as: high/low availability of alternative antecedents, and as high/low availability of disabling conditions. These were then administered with the conditional inference task to a separate group of participants. Like Hilton *et al.* (1990) they did not instruct participants to evaluate the necessity of the present conclusions, and in fact allowed a graded response on a seven-point scale from 'very sure I *cannot* draw this conclusion' to 'very sure that I *can* draw this conclusion'. The results were very clear cut. For conditionals with many disabling conditions, the endorsement of conclusions for the valid inferences MP and MT was substantially and significantly lower than for conditionals with few disabling conditions. The suppression of the 'obvious' MP inference is particularly striking. They also found a significant, but smaller effect of alternative antecedents. As expected, and consistent with earlier studies, conditionals with more alternative antecedents led to fewer DA and AC inferences. Like most authors in this field,

Cummins *et al.* (1991) discuss their results in terms of conditional and biconditional readings. We prefer to talk of the sufficiency and necessity relations. The point is that in context either of these relations can be strong or weak, which leaves four possible categories not two. If sufficiency is weak then neither the conditional nor biconditional reading applies. The findings of Cummins *et al.* have recently been replicated in essential respects by De Neys *et al.* (2003a).

While striking, the findings of Cummins *et al.* leave open a couple of important questions. Is the effect restricted to 'causal' conditionals (those uttered in a causal context), and would the findings hold up if genuine deductive reasoning instructions were employed? The answers to both were supplied in a study reported by Thompson (1994). Starting with a large pool of thematic conditionals, she asked one group of participants to rate each for the degree to which p was sufficient for q, and the degree to which p was necessary for q. A second group classified the sentences into four contextual categories: obligation, permission, causal, or definition. A third group was then given conditional inference problems with sentences of all four contextual categories. Within each context, there was a further subdivision of four types depending upon high/low sufficiency and high/low necessity. The instructions used in this study were strict deductive reasoning instructions. Participants were told to answer the questions only on the basis of answers that followed logically from the conditional statement and were only allowed a choice of yes, not, and maybe in evaluating conclusions.

Thompson's findings were complicated by an interaction between sufficiency and necessity, but simplifying a little, the results conformed with predictions. Valid inferences, MP and MT were made less often when sufficiency was lower, and fallacious inferences, DA and AC, were less frequently endorsed when necessity was lower (see Figure 6.1). There was also an unexplained increase in valid inferences under high perceived necessity (see Figure 6.1b), although this effect was much smaller than for the fallacies. These effects were observed for *all four* contextual categories, whose data are combined in Figure 6.1. The major effect of using deductive reasoning instructions was to reduce the overall frequency of fallacies, which were made substantially less often than valid inferences in this study. Hence, the findings are very similar to those in the belief bias literature reviewed earlier and compatible with dual process theory. Deductive reasoning instructions appear to induce an effort at deduction, at least in some of the participants, which leads them to make good overall discrimination of valid from invalid inferences. However, such instructions do not remove the pragmatic influence of prior belief and knowledge, which also substantially affect the responses given to the same reasoning task.

In Chapter 5, we saw that the deontic selection task was a good deal easier than the indicative one, although the comparison is somewhat confounded by the fact that most indicative selection tasks have used abstract content and most deontic tasks realistic content. Nevertheless, the reader might be surprised by the finding of Thompson (1994) that perceived necessity and sufficiency has much the same effect on conditional inference for deontic contexts (permission and obligation) as for causal and definitional conditionals. In fact, in more recent work, Thompson (2000) has shown that context effects are task specific. Her research indicated that deontic context predicts superior performance on the selection task but not on the conditional inference task. Conversely, she also showed that perceived necessity and sufficiency predicts inference task and not

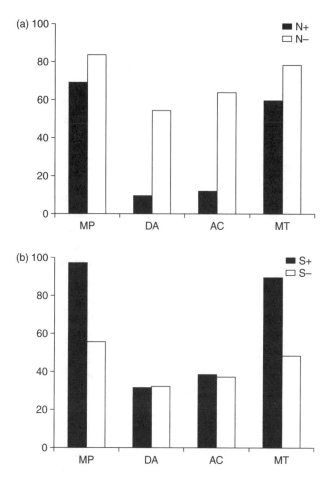

**Figure 6.1.** Conditional inference rates in the study of Thompson (1994). (a) Effect of high (N+) and low (N−) perceived necessity. (b) Effect of high (S+) and low (S−) perceived sufficiency.

selection task performance (but see Ahn and Graham, 1999). We see this as evidence for the account of the selection task in terms of relevance (Evans, 1989) and of the extension of this to a decision theoretic account (Evans and Over, 1996a, see chapter 6). Most people do not use deductive reasoning in the selection task, but rather make decisions based on perceived relevance and expected benefits and costs.

The results reviewed in this section are representative and known to be highly reliable. Similar findings have been reported in various other studies in the literature, including a set of studies of Markovits and colleagues, which include study of children as well as adults (Markovits, 1984, 1986; Markovits *et al.*, 1998; Quinn and Markovits, 1998, 2002; Barrouillet *et al.*, 2001). This has led Markovits to the view that the ability to access relevant information from semantic memory—in this case alternative antecedent or disabling conditions—develops as children grow older. Whether this is seen as paradoxical or not depends upon your view of the deduction paradigm (Evans, 2002a).

If you see the influence of prior belief in a deductive reasoning task as a bias, then it may seem paradoxical that such belief biases 'develop'. If, on the other hand, you see the ability to retrieve and apply relevant knowledge to the task in hand to be a generally adaptive feature of everyday cognition (Evans and Over, 1996a) then, of course, it makes a good deal of sense. Markovits and colleagues (for example, Quinn and Markovits, 1998; see also De Neys *et al.*, 2003) have also shown that it is not just the number of alternative antecedents or disabling conditions that can be retrieved from memory that determines the effects on reasoning, but also their degree of associative strength.

Before leaving this section we should mention the case of only-if conditionals that have been mentioned earlier in this book (see Chapters 1 and 3). We earlier reviewed some evidence for a directionality effect. People prefer to reason forwards (MP, DA) with if-then conditionals, but backwards (AC, MT) with only-if conditionals, a pattern that defies an account in terms of truth tables. Evans (1977) suggested that with deontic conditionals we may prefer the if-then form when we want to emphasize the necessity of the consequent rather than the sufficiency of the antecedent. Consider the following examples of logically equivalent statements:

6.4   If you drink in a bar then you must be over 18 years of age
6.5   You may drink in a bar only if you are over 18 years of age

Unlike with temporal-causal conditionals both if-then and only-if forms are linguistically acceptable but seem subtly different in meaning. Statement 6.5 seems to give more emphasis to the fact that it is necessary to be over 18 in order to drink in a bar, which may well be why participants generally make more MT inferences with only-if conditionals, even when framed with abstract content. Thompson and Mann (1995) investigated thematic conditionals that were either deontic (permission) or arbitrary in nature. Participants were asked to rate both the necessity of p for q and the necessity of q for p. They were also asked to judge the semantic similarity of both 'p only if q' and 'q only if p' to an original conditional 'if p then q' for a number of different thematic conditionals. In permission contexts, such as those illustrated by 6.4 and 6.5 above, participants strongly endorsed 'p only if q' as being similar to 'if p then q' and also rated q as being highly necessary for p (rather than the other way around). With arbitrary conditionals, there was no clear preference for either paraphrase of the conditional and correspondingly no clear preference for the direction of necessity. These findings show clearly that 'p only if q' does, as Evans (1977) suggested, convey the necessity of q for p, but does so most clearly in deontic contexts.

We have already noted that most researchers on this topic refer to conditional and biconditional interpretations of the conditional. There has also been a trend towards application of the mental model theory of Johnson-Laird and Byrne (1991) or variants on it in explanation of these pragmatic effects (for example, Thompson and Mann, 1995; Markovits and Barrouillet, 2002; De Neys *et al.*, 2003). However, these references often talk of retrieval of FT cases as *counterexamples* that suppress fallacies (for recent examples, see Markovits, 2002; De Neys *et al.*, 2003). It is important to note that the mental model theory of conditionals presented by Johnson-Laird and Byrne (2002) appears to *preclude* the representation of false cases, in line with the 'principle of truth'. What they actually say (p. 654) is, 'According to the principle of truth, mental models represent true possibilities. The principle does not imply, however, that individuals never represent false cases. They can use their mental footnotes about what is false to construct fully explicit models of what is true.' We find this statement rather confusing, not least because it is hard to know what kind of representation a 'mental footnote' might be. However, we will do our best to apply the theory.

As we explained in Chapter 4, the only clear explanatory mechanism offered by Johnson-Laird and Byrne (2002) for the pragmatic modulation of mental models is what we term the 'four-bit device'. The effect of pragmatic modulation is to modify the list of possibilities in the light of real world knowledge. How could this account for effects of perceived necessity and possibility? In accordance with their principle of truth, only true possibilities are explicitly represented in mental models. Hence, the knowledge (mental footnote?) that FT is forbidden in some contexts (strong necessity of p for q) must be introduced indirectly. The initial representation of just the TT case can be fleshed-out according to context and task demands. For a basic conditional, this fleshing-out would result in the addition of FT and FF as true possibilities, sanctioning MT as well as MP, but not the fallacies. However, in a biconditional context where FT is blocked by pragmatic knowledge, participants would flesh-out only the FF case as an additional possibility to TT. This explicit model set allows both AC and DA inferences to be sanctioned.

This mental models account of the necessity manipulation is still extensional, that is described simply in terms of the set of logical possibilities that are allowed. It seems to us to add little in the way of explanation to these effects and is equivalent to a truth functional chameleon theory in which representation switches between *material* conditional and biconditional readings. A pragmatic modulation account of sufficiency effects—weak necessity inhibits the valid inferences MP and MT—would seem to have to be that the conditional is represented as a tautology in which TF is added to the extension of possible cases, so that everything is possible, and none of the inferences can be drawn. However, research on disabling conditions shows that it is possible to undermine the relation 'p is sufficient for q', while leaving the relation 'p is necessary for q' intact. Presumably, this is represented as a reverse conditional, 'if q then p', which permits everything except FT. Pragmatic modulation taken to this extent seems to leave little trace of the original mental representation of conditionals proposed in the theory.

## Suppression of conditional inferences

In the above section, we have reviewed studies that relate people's prior belief about conditional statements to the rates with which they will endorse conditional inferences. Essentially this showed that where alternative antecedents can easily be retrieved from memory people are less inclined to make the fallacies DA and AC; similarly, where disabling conditions are available they are less inclined to make the valid MP and MT inferences. We could think of these retrieved memories as 'suppressing' inferences. However, the term *suppression* has been associated by psychologists with a different paradigm in which conditional inferences are suppressed by an active intervention in which additional premises are presented. We could also view these as studies of defeasible conditional inference.

The paradigm was first introduced by Byrne (1989; Byrne, 1991). Consider the following MP problem:

> If she meets her friend she will go to a play
> She meets her friend
> Therefore, she will go to a play

96% of participants endorsed this as a valid inference. However, if we now add a second conditional premise:

> If she meets her friend she will go to a play
> If she has enough money she will go to a play
> She meets her friend
> Therefore, she will go to a play

endorsement rates drop to just 38%. A similar manipulation with MT (minor premise: she does not go to a play, conclusion: she does not meet her friend) dropped from an endorsement rate of 92–33% when the second conditional was added.

Byrne's mental model account of this finding, as recently developed by Byrne *et al.* (1999), is as follows. It is essentially the counterexample argument that we discussed in the previous section as the generalized mental model account of sufficiency/necessity effects. When the two conditionals are presented, people are proposed to form an extended model that incorporates both:

> friend   money   play
> . . .

and that makes available a counterexample to the valid MP inference from friend to play. We find the above mental models vague as they might represent any statements that allow this model plus alternatives to it. For example, this could be:

> If she meets her friend and has money then she goes to the play

as Byrne suggests, which suppresses MP. However, why could this model set not equally represent:

> If she meets her friend then she has money and goes to the play

which would permit the original MP inference? If this cannot be allowed, then either there is something *additional* to the mental models that Byrne and colleagues describe that represents these sentences, or the models they give are incomplete. Either way it is unsatisfactory. This is very important, because it is the claim that pragmatic factors can be accounted for simply in terms of the logical possibilities that people consider that is the cause of our main doubt with the model theory. At best, it seems to us that the pragmatic side of the theory is highly underspecified.

The model account of the suppression of valid conditional inferences was challenged early on by mental rule theorists (Politzer and Braine, 1991), who argued that the effect of the second conditional is to undermine belief in the first. People resist MP (and MT) because they seriously doubt the conditional from which this inference derives. Recall that in our review of the belief bias literature, we found evidence that despite instructions to judge the validity of conclusions, participants in reasoning experiments are inclined to judge their soundness instead. If this is right, then we could certainly expect people to resist making inferences from premises that that they disbelieve. This line of argument has been developed by other authors who are not sympathetic to the mental rule theory, but prefer a probabilistic account of conditionals, about which we shall have much more to say in Chapter 8. Stevenson and Over (1995), for example, discuss suppression of valid inferences by the following premises:

If John goes fishing, then he will have a fish supper
If John catches a fish, then he will have a fish supper

The second conditional, when present, will tend to suppress MP and MT inferences drawn from the first. What, however, if some further qualifying conditions are added from the following choices:

Q1—John is always lucky when he goes fishing
Q2—John is nearly always lucky when he goes fishing
Q3—John is sometimes lucky when goes fishing
Q4—John is rarely lucky when he goes fishing
Q5—John is very rarely luck when he goes fishing

Stevenson and Over (1995; Experiment 1) told participants to assume that the statements given were true and to choose the conclusion that they felt followed. They presented their problems as pragmatic rather than deductive tasks and asked people to choose from five statements to either an MP or MT problem, expressing various degrees of certainty in the deductive valid conclusion, one of which was the valid conclusion that the relevant event happens. Both choice rates of the valid conclusion and mean confidence scores computed across all response choices progressively declined from Q1 to Q5 depending upon which qualifying statement accompanied the two conditionals, despite the instruction to assume the truth of the premises (see Figure 6.2). In a second experiment they used pragmatic reasoning instructions, telling people to imagine they were listening to a conversation and to indicate what they thought followed. The same trends were observed, but the proportion of certain or determinate conclusions dropped sharply from that observed in the first experiment. Once again, we see evidence that use of deductive reasoning instructions causes an effort at deduction, which modifies but does not remove the effect of pragmatic factors.

The findings and conclusions of Stevenson and Over have been supported in a series of papers by George who has shown a strong relation between the believability of the conditional premise and the confidence with which people will draw conditional inferences. Instead of presenting additional premises, as in the experiments we have just been discussing, George (1995) looked at the prior believability of particular sentences, a methodology similar to that used to study perceived sufficiency and necessity effects. In fact, it is probable that rating conditionals for sufficiency and for believability would produce similar findings. In a within-participant design, George (1995, Experiment 1) showed that people's willingness to endorse MP for particular conditionals was strongly correlated with their believability ratings of the same conditionals in a separate task. This occurred in spite of strong instructions to assume the truth of the premises in the conditional inference task.

In a later experiment, George (1995) compared the effect of such instructions with another set, which told people to take into account the believability of the premises.

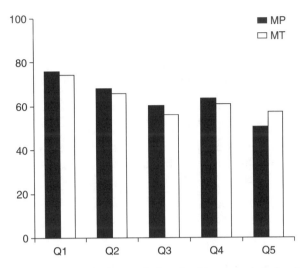

**Figure 6.2.** Percentage confidence in the conclusion of valid arguments in the study of Stevenson and Over (1996, Experiment 1). Note: The data were reported originally as uncertainty scores on a scale from 1 to 5. These have been transformed by subtracting from 5 and multiplying by 20 to provide estimates of percentage confidence.

This produced a marked shift with responding to more strongly belief-based responding. George suggests that in his first experiment some participants were assumption based (logical) and some belief-based in their approach, but that the change in the instruction caused everyone to shift to the latter approach. This conforms with our previous observations about belief bias and dual process theory. We should also note that Stanovich (1999) has found evidence that high ability participants are better able to comply with deductive reasoning instructions and suppress pragmatic influences. The instructional effects in both George's and Stevenson and Over's study could then be interpreted on the hypothesis that the deductive instructions were successful with only some participants, hence reducing but not eliminating the effects of prior belief.

We will return to the issue of the believability of conditionals and its relation to probability in Chapter 8. For now, we note that the interpretation of the suppression of valid inferences offered by Poltizer and Braine (1991) is strongly supported by the evidence, to the extent that people do resist inferences when they disbelieve the conditional. Stevenson and Over (1995) also reversed the effect by applying a qualifying statement in some cases that restored faith in the original conditional. Bonnefon and Hilton (2002) have used this as evidence for their own somewhat different account of the suppression effect based on conversational implicatures. Given the two conditionals:

If John goes fishing, then he will have a fish supper
If John catches a fish, then he will have a fish supper

they argue that there is a clear implicature (an inference that the speaker intends the listener to draw) that John may not catch a fish and thus be unable to have a fish supper. People do not just think of the counterexample case: they are positively invited by the speaker so to do. If the speaker adds 'John is always lucky when he goes fishing' then this is equally clearly intended to cancel the above implicature. This paper conforms with Hilton's general approach, which is to explain the results of reasoning experiments on the grounds that these are social interactions between experimenter and participant to which Gricean rules of conversation apply (Hilton, 1995).

Stevenson and Over (2001) showed that valid inferences can also be suppressed by making the minor premises of conditional inferences uncertain. This result cannot be explained by the mental model account of Bryne (1989, 1991) and Byrne *et al.* (1999). In their experiments, Stevenson and Over presented premises as if these were statements by people with different levels of authority in a dialogue. The participants responded by showing that they had more confidence in the conclusions when high authority, expert figures stated the premises than when low authority, novice figures did. This response was present whether the high or low authority figures stated the major or the minor premises. For example, compare these two sets of premises:

*Professor of Medicine*: If Bill has typhoid then he will make a good recovery
*Situation*: Bill has typhoid

*Medical student*: If Bill has typhoid then he will make a good recovery
*Situation*: Bill has typhoid

In the above example, the major premise, for MP, is stated by either an expert or by a novice, and the minor premise is given as a fact in both cases. The participants had more confidence in the conclusion, that Bill will make a good recovery, when the expert stated the major premise. This result was also found when a comparison was made between the statement of the minor premise by an expert or a novice, as in:

*Rule*: If Bill has typhoid then he will make a good recovery
*Professor of Medicine*: Bill has typhoid

*Rule*: If Bill has typhoid then he will make a good recovery
*Medical student*: Bill has typhoid

In this second example, the major premise is given as a fact, and the minor premise is stated by either an expert or a novice. The participants expressed more confidence in the conclusion of MP when the expert asserted the minor premise than when the novice did. From a normative point of view, uncertainty in the minor premise should have an effect on confidence in the conclusion, just as uncertainty in the major premise does.

Stevenson and Over (2001) also studied premises stated by figures with different levels of expertise. For example, the first conditional, or major, premise in the following might the stated by a student and the second conditional premise by a Professor of Medicine:

If Bill has typhoid then he will make a good recovery
If Bill has malaria then he will make a good recovery
Bill has typhoid

Similarly, a student might assert the first categorical, or minor, premise below and a Professor of Medicine the second categorical premise:

If Bill has typhoid then he will make a good recovery
Bill has typhoid
Bill has malaria

The degree of confidence the participants had in the conclusion of MP depended on the level of expertise of the speakers who asserted the major or minor premises and the second premises in each instance. There was more suppression of MP when a novice asserted the first premise and an expert the second than when an expert asserted the first premise and a novice the second. This effect was found for both major premises with a second conditional premise and minor premises with a second categorical premise. It confirms the view that suppression is caused by uncertainty in the premises, whether this uncertain is in major or minor premises.

## Inducement and advice conditionals

A special category of conditional statements are those offered by a speaker with the intention of changing the behaviour of the listener by offering advice or inducements. Specifically, these fall into four categories, illustrated by examples as shown below:

*Promise*: father to son
If you pass the exam, I will buy you a bike

*Tip*: friend to friend
If you pass the exam, your father will buy you a bike

*Threat*: boss to employee
If you are late for work again, I will fire you

*Warning*: colleague to colleague
If you are late for work again, the boss will fire you

Pioneering work on conditionals of this type is reported by Fillenbaum (1975, 1976). Fillenbaum noted that such conditionals seem strongly to invite the DA, not-p therefore not-q, and demonstrated this with experiments using pragmatic reasoning instructions. More complete data on reasoning with these pragmatic types are provided by Newstead *et al.* (1997). Their experiments presented conditionals in short conversational contexts as in the following example of a warning conditional (p. 56):

Sandy, a staunch Everton supporter, was discussing with his father whether it was safe to wear his team's colours when travelling to the forthcoming game with Arsenal. Sandy's father thought not, and told him:
'If you wear Everton's colours to the match you'll be beaten up on the train.'

Newstead *et al.* carried out both truth table tasks and conditional inference tasks on a range of pragmatic types of conditionals including the four on which we focus here. In the truth table task they were given the four contingencies:

Sandy did wear the Everton colours to the match:     TT
He was beaten up on the train

Sandy did wear the Everton colours to the match:     TF
He was not beaten up on the train

Sandy did not wear the Everton colours to the match:   FT
He was beaten up on the train

Sandy did not wear the Everton colours to the match:   FF
He was not beaten up on the train

Participants were asked to say whether each outcome 'supports' the conditional sentence, 'contradicts' it or 'tells us nothing about it'. In some respects, the findings resembled those for truth table tasks with abstract indicative conditionals, discussed in Chapter 3. TT cases were almost universally said to support the statement and TF cases to contradict it. There were also frequent classifications of FT and FF cases as 'telling us nothing' about the statement, in line with the defective truth table. However, the data of most interest were responses that indicated FT as contradictory, because these suggest a biconditional reading. These were common (Experiment 1) for promises (56%) and threats (50%) but less common for tips (26%) and warnings (30%). This appears to suggest the latter two types are less biconditional in interpretation. Another finding of interest was that FF cases were quite often described as supporting the conditional (as opposed to being irrelevant): this occurred in Experiment 1 for 68% of promises, 70% of threats, 51% of tips and 58% of warnings. Note that the weaker forms—tips and warnings—of these conditionals again produced lower ratings.

When individual patterns were analysed it transpired that many individual participants classified FT as irrelevant but FF as true for these types of conditional—a previously unnoticed pattern. Here is a possible account based on the T3 view of conditionals. Suppose people apply the Ramsey test to the original affirmative conditional:

If you wear Everton's colours to the match, you'll be beaten up on the train

They easily imagine a possibility in which the colours are worn and they are beaten up (TT) and see this as supporting the statement. The TF statement is then seen as contradicting this state of affairs (in the antecedent state of affairs, the consequent should be true, not false). Now suppose that the effect of pragmatic implicature is to add the second conditional, 'if not p then not q':

If you do not wear Everton's colours to the match you'll not be beaten up on the train

The Ramsey test for this added conditional leads you to imagine a possibility in which you do not wear the colours and are not beaten up. But of course this is the FF case of the original conditional and may thus account for the high ratings of FF as true in this experiment. On the FT cases, however, participants divide between judgements of false and irrelevant. This could be due to a problem with double negation. This possibility is one in which you do not wear the colours and do *not* get beaten up. If participants fail to process this double negation then they may say that FT is irrelevant.

T3 plus pragmatic implicature is a possible way to explain the finding, as two conditionals, each with a defective truth table, are combined in the representation. However, there is also a possible explanation based on the T2 view of conditionals. Recall that the difference is that with the T2 conditional people judge the truth of a conditional in false antecedent cases by reference to the *closest* possible state of the world in which the antecedent holds (Chapter 2). In this case, the effect of inverse implicature may be to draw participants' attention to FF as a close possibility, seen as a confirming what the speaker meant to convey. No explicit second conditional is generated in this account. We again have the problem of explaining why FT is seen as irrelevant rather than false. Perhaps there is a closeness heuristic that cues the relevance of close possibilities. This is analogous to the effects of matching bias on the abstract truth table task (Chapter 3) in which mismatching cases are discarded as irrelevant.

Table 6.2. Conditional inference rates for inducement and advice conditionals

|  | MP | DA | AC | MT |
|---|---|---|---|---|
| (a) Newstead Ellis and Dennis (1997, Experiment 4) | | | | |
| Promise | 89 | 74 | 78 | 73 |
| Tip | 60 | 40 | 41 | 38 |
| Threat | 93 | 84 | 71 | 84 |
| Warning | 75 | 55 | 55 | 64 |
| (b) Evans and Twyman-Musgrove (1998) | | | | |
| Promise-HC | 82 | 66 | 49 | 42 |
| Promise-LC | 61 | 55 | 37 | 25 |
| Threat-HC | 82 | 75 | 34 | 38 |
| Threat-LC | 62 | 60 | 29 | 33 |
| Tip-HC | 77 | 61 | 58 | 37 |
| Tip-LC | 48 | 38 | 27 | 23 |
| Warning-HC | 83 | 64 | 47 | 61 |
| Warning-LC | 51 | 43 | 31 | 25 |

HC, high perceived control by speaker over consequent event; LC, low perceived control by speaker over consequent event

Newstead *et al.* (1997, Experiment 4) also investigated conditional inferences for the same pragmatic types. Sentences were again presented in conversational context and participants simply asked to decide whether conclusions followed. There was no reference to logic, assuming the premises, or drawing necessary conclusions. Hence, this should be viewed as a pragmatic rather than deductive reasoning task. The frequencies of inferences are shown in Table 6.2(a). As Fillenbaum predicted, DA inferences are very common for promises and threats. Indeed, all four conditional inferences are frequently made with these types. However, inference rates are substantially lower for tips and warnings. What is interesting is that this trend applies across *all four* inference types. There is some sense in which promises are stronger than tips, and threats are stronger than warnings.

Newstead *et al.* suggested that promises and threats are stronger because the speaker has control over the consequent event. Two subsequent studies have confirmed this hypothesis using different methodology. Evans and Twyman-Musgrove managed to produce high and low speaker control versions of each pragmatic type and showed decisively that speaker control was the factor determining inference rates. Their data, shown in Table 6.2(b) also showed that the effect of low speaker control is to reduce the rates across all four inferences. Dieussaert *et al.* (2002) using a range of thematic conditionals and a multiple regression approach showed that speaker control was a strong predictor of perceived likelihood of both the consequent given the antecedent and vice versa.

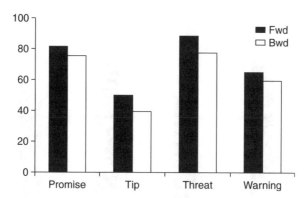

**Figure 6.3.** Comparative rates of forward (MP and DA) and backward (AC and MT) inferences made in the study of Newstead *et al.* (1997).

It appears then that we evaluate the credibility of those who assert conditionals and our willingness to draw any inference depends upon this. This conclusion is also supported by the studies of Stevenson and Over (2001) that we reviewed in the previous section. Once again, we find a link with the idea of sound inferences discussed when reviewing the belief bias effect. Why is it that this should affect all four conditional inferences? We believe—as suggested above—that people are adding, by pragmatic implicature, the inverse conditional 'if not p then not q' to the presented 'if p then q' and reasoning with both. This generates MP and DA inferences directly. There may well be a second, but weaker, invited converse inference for each conditional that can produce AC and DA but only if each conditional is credible. Thus we predict a directional bias in the reasoning: people will give more forward inferences (MP, DA) than backward inferences (AC, MT) as a direct consequence of applying the Ramsey test to both original and inverse conditionals.

Inspection of Table 6.2 confirms that such forward reasoning is indeed stronger. [In the study of Evans and Twyman-Musgrove (but not that of Newstead *et al.*) the trend is clearly shown if only fallacious inferences are considered. In this study, DA was endorsed overall 58% of the time, but AC only 39% of the time.] For clarity, we have summarized the relevant data in Figures 6.3 and 6.4 where the trend emerges very clearly. Note that this does not fit with the chameleon theory, favoured in the literature on perceived sufficiency and necessity. Conditional versus biconditional readings yoke together MP and MT on the one hand, and DA and AC on the other. This could not produce a yoking of MP and DA as opposed to AC and MT as actually occurs in these experiments. In fact, we do not believe this trend can be accounted for in terms of the representation of logical possibilities at all. Hence, it is not explicable within our extensional interpretation of what Johnson-Laird and Byrne (2002) call 'pragmatic modulation' (see Chapter 4). Although Johnson-Laird and Byrne's (2002) allude briefly to directionality in mental models, they have not to our knowledge proposed an amendment to their extensional reasoning mechanism to explain how this can actually affect the inferences drawn. It certainly involves more than simply drawing inferences from permitted possibilities.

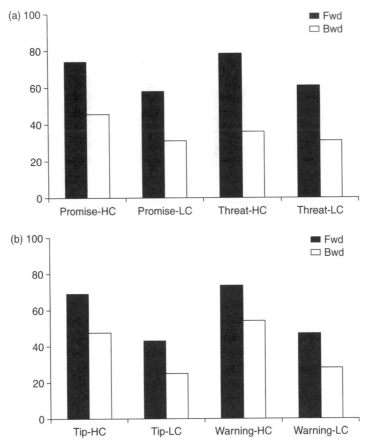

**Figure 6.4.** Comparative rates of forward (MP and DA) and backward (AC and MT) inferences made in the study of Evans and Twyman-Musgrove (1998). (a) Promise and tips. (b) Threats and warnings.

Related to this point, the research of Newstead *et al.* (1997) shows the limitation of the truth table method based upon asking participants to classify logical possibilities. It is evident that this method does not capture everything about the mental representation of conditionals in different pragmatic contexts. First, there is nothing in the truth table data that would allow us to predict the observed preference for forward over backward inferences. Second, the truth table data only hint at the stronger nature of promises and threats as compared with tips and warnings. Recall that ratings of FT as false and of FF as true were substantially stronger for the former pair than the latter. We can see a rationale for this, by application of the Ramsey test to the added inverse conditional as already explained. We could not, however, have predicted the across the board drop in inference rates for tips and warnings as their rates of evaluation of TT as true and TF as false were as strong as for promises and threats.

# Conclusions

There are some clear conclusions to be drawn from the studies discussed in this chapter. First, it is evident that people are strongly influenced by prior belief and relevant knowledge when reasoning with thematic conditionals, even when clear deductive reasoning instructions are employed. The main influence of such instructions is to reduce (but not eliminate) conditional 'fallacies', presumably due to the emphasis on necessity. Instructing people to assume the premises and/or base their reasoning on the given information reduces to some extent the influence of prior belief, but by no means eliminates it. We believe this because pragmatic reasoning is the strong default mode that we use in everyday life and for very good reason. It nearly always serves our goals to reason from *all relevant belief*, even though Stanovich (1999) has labelled our compulsion to contextualize the 'fundamental computational bias'. When instructed to reason deductively, we find this very difficult and despite making an effort at deduction, few participants, other than those with very high general intelligence, are able to resist the influence of prior belief or consistently find solutions requiring abstract logical reasoning.

When we are reasoning about 'causal' conditionals, in particular, our prior knowledge influences the perception of both the sufficiency of p for q and of its necessity. We will not draw the valid MP and MT inferences if we can think of disabling conditions that would prevent p leading to q. Similarly, we will not draw the DA and AC inferences if we can easily imagine alternative causes for q other than p. Similarly, effects of perceived sufficiency and necessity have been shown with other pragmatic types, including deontic conditionals. The literature on suppression of conditional inferences (or defeasible inference) shows the same effects by a different method. Instead of relying on people to think of counterexamples cases for themselves, additional premises are presented that suggest the presence of additional antecedents or disabling conditions, with similar effects.

We believe that the almost universal account of these studies in terms of conditional and biconditional representations to be oversimplified, and that the way in which many authors refer to the mental model theory of Johnson-Laird and Byrne essentially does no more than restate the chameleon theory in which a conditional statement is read as a material conditional or material biconditional according to context. We also believe that the model theory has serious problems in accounting for why DA is often made more than AC with thematic conditionals. The preference for AC inferences over DA with abstract conditionals (see Chapter 3) is explicable in the theory because AC could be made from the initial representation, whereas DA requires fleshing-out. We know of nothing in the mental model theory of thematically rich conditionals that could account for a reversal of this trend. Under our account, in which the conditional is not-truth functional, converse and inverse implicatures need not go together (see Chapter 9 for further discussion of this point).

The research on inducement and advice conditionals is particularly interesting as it provides inference patterns that cannot be explained in terms of conditional/biconditional readings nor in terms of pragmatic modulation of mental models representing logical possibilities. People make all inferences more commonly with the stronger promises and threats than with the weaker tips and warnings, something we know to be linked to perceived control of the consequent by the speaker. They also make more forward inferences (MP, DA) and than backward inferences (AC, MT) with conditionals in these pragmatic categories. We have suggested an account that combines the Ramsey test with the use of pragmatic implicature. A more complete account of our theory is presented in Chapter 9.

We should note finally that a number of the studies reviewed in this chapter suggest that the representation of thematic conditionals and the inferences drawn may be probabilistic. It is true that the majority of studies ask people to make yes or no decisions, and that the proportion of participants giving one answer or the other is what typically relates the variables manipulated. However, we also looked at several studies where participants were allowed to express a degree of confidence in the conclusion. In general, the two measures seem to coincide. As the proportion of people endorsing conclusions drops by the first method, so the average rating of confidence drops by the second. In Chapter 8, we will look at some studies specifically designed to test the idea of probabilistic conditional reasoning and also revisit some of the research discussed here in that context.

# 7 Counterfactuals: philosophical and psychological difficulties

We have mostly concentrated in this book on indicative conditionals, although we have also discussed deontic conditionals in Chapter 5. Another important category of conditional statements are known as *counterfactuals*. These conditionals are often said to be in the subjunctive mood, but use of this mood is neither necessary nor sufficient to make a conditional a counterfactual (Dudman, 1988; Edgington, 1997; Woods, 1997). On the other side, there are some good reasons for using 'subjunctive conditional' in place of 'counterfactual' as a technical term (Bennett, 2003). However, we will follow the normal psychological practice of using 'counterfactual'. The study of counterfactuals has resulted in a rich philosophical and psychological literature, some of which we will refer to below. More recently, techniques from computer science and AI have been applied to this study (Pearl, 2000; Wobcke, 2000). Counterfactual thinking is clearly a form of hypothetical thinking and as such should eventually be encompassed within the broad theoretical approach to conditionals that we are advocating.

No one, or almost no one, has suggested T1 for counterfactuals, but there are versions of both the T2 and T3 families for counterfactuals (Edgington, 1995; Bennett, 2003). As a reminder (of Chapter 2), both T2 and T3 avoid the paradoxes of the material (T1) conditional but differ in their treatment of false antecedent cases. There is a truth value gap for Adams, or T3, conditionals in a false antecedent cases, similar to the defective truth table discussed in Chapter 3. But there are always truth values for Stalnaker, or T2, conditionals in false antecedent cases, depending on what holds in the closest possibilities in which the antecedents are true. Whatever the normative position, psychological research is a long way, in our opinion, from determining whether T2 or T3 gives the better descriptive account of counterfactuals.

Let us begin by illustrating what has been taken to be an important difference between indicative and counterfactual conditionals. We will adapt an example from Chapter 2 and suppose that a doctor usually keeps only two kinds of test on his desk: a test for diabetes and a test for pregnancy. The doctor takes a test from his desk, gives it to a woman patient, and the result goes off to a laboratory to be analysed. The result comes back positive, but the doctor is absentminded and cannot recall why the woman came to see him or what kind of test he gave her. On general grounds, and with knowledge of his past behaviour, he is confident that the test was either for diabetes or pregnancy. He is therefore confident that:

7.1   The patient had diabetes or she was pregnant

From the disjunction, 7.1, the doctor can justifiably infer the indicative conditional:

7.2   If the patient did not have diabetes, then she was pregnant

However, many philosophical logicians would argue that the doctor would not be justified in inferring from 7.1 the counterfactual:

7.3   If the patient had not had diabetes, then she would have been pregnant

Notice that the indicative 7.2 follows from the disjunction 7.1 only in special circumstances. A more competent, second doctor might know that one of her patients had diabetes, in which case she can validly infer 7.1 by or-introduction. But she would not infer 7.2 from 7.1, as the result would be an absurdity essentially like P1, a 'paradox' we met in Chapter 2. To use a technical term, this second doctor has a *constructive* justification of 7.1, which is in the form 'p or q'. That means that she not only believes 'p or q', but she can identify whether p holds or q holds (or both hold). The first doctor has a *non-constructive* justification of 7.1. He believes 'p or q' on general grounds, but does not know which proposition, p or q, holds. In his epistemic state, he can justifiably infer 7.2 from 7.1. He can infer, unlike the second doctor, that 7.2 is probable given 7.1. The inference from 7.1 to 7.2 is therefore not logically valid for ordinary conditionals, but sometimes is, and sometimes is not, a reasonable inference. It can be seen to be reasonable by the Ramsey test or in Stalnaker's analysis (Stalnaker, 1975), depending on one's subjective epistemic state. There are interesting psychological experiments on this type of inference (Ormerod and Richardson, 2003) but unfortunately they do not distinguish between validity and reasonableness.

We should stress that having a non-constructive justification is being in a certain epistemic state. We can justifiably believe that the woman patient has diabetes or is pregnant, without knowing which. Obviously, however, the objective state of affairs cannot be such that the woman has diabetes or is pregnant, but it is not determined which. These points suggest to some philosophers that at least some indicative conditionals, like 7.2, are epistemic in essential nature and used to express epistemic states (Bennett, 2003).

Many philosophers would say that the counterfactual 7.3 does not follow from 7.1, not even as reasonable inference. This point has been seen to support the conclusion that counterfactuals are to do with objective, non-epistemic states, such as causal or other objective relations. Whether we have a constructive or a non-constructive justification of 7.1, we are not inclined to infer 7.3 from 7.1. We do not see how the patient's not having diabetes could objectively have had anything to do with her being pregnant. The difference between apparently epistemic indicative conditionals, and non-epistemic counterfactuals, has led some philosophers to argue for T3 for indicatives but T2 for counterfactuals (Gibbard, 1981; Bennett, 2003).

Counterfactuals may, or may not, be more about objective relations than at least some indicative conditionals, but they still have to be processed psychologically. Consider an example of how this might be done. Suppose that we are reviewing the moves we recorded, in a booklet, of a chess game that we lost some time ago. We suddenly notice that we missed a quality move at one point. We think to ourselves, or assert to a friend who is going over the game with us, a conditional:

7.4   If I had taken that pawn, then I would have won the game

This conditional is a typical counterfactual. In this context, it is a pragmatic presupposition, and common knowledge, that we have already lost the game. We are talking about a state of affairs, in which we win the game, that was once possible, but is no longer. The grammar of 7.4 may help to suggest that it should be interpreted as a counterfactual, but context is what decides the matter. One mark of a counterfactual is a presupposition or suggestion, in a particular context, that the antecedent of the conditional is false or at least probably false.

We might have very high confidence in 7.4 after we work through, in retrospect, all the counter moves our opponent might have made in response to our taking his pawn. Doing that, we might conclude that he had little chance of making an effective counter move. In that case, our attitude to the antecedent of 7.4 could be important. Perhaps capturing our opponent's pawn is just the sort of move we often make in positions like the one we were in at that point of the game. We might even have considered taking the pawn, but failed to do so because of uncharacteristically hesitant play. Perhaps we nearly took the pawn, but then chose a seemingly safer move for a reason that not even we found all that compelling at the time. In cases like this, we would probably feel a great deal of regret that we did not capture the pawn. We would be unhappy about losing the game and then feel even worse because we could so easily have won it and were so close to doing so. This whole process of studying the game and working out the moves in it we might have made, and our consequent feelings about our loss, could help us to improve our decision making in future games. We will certainly be far less likely to fail to take a piece in similar circumstances in the future.

A full psychological theory of counterfactuals should account for the cognitive processes that produce and justify counterfactuals, the emotional reactions that often accompany these processes, and the subsequent effect on decision making. In our view, some work in philosophical logic (especially that following on from Stalnaker, 1968) is highly relevant to accounting for the cognitive processing. But also relevant are studies of counterfactual thought by social psychologists and psychologists working on judgement and decision making (especially that following on from Kahneman and Tversky, 1982; Kahneman and Miller, 1986). Psychologists studying reasoning have proposed theories of the cognitive processes (Rips and Marcus, 1977; Braine and O'Brien, 1991; Johnson-Laird and Byrne, 1991, 2002). These theories are surprisingly poorly integrated with the work in philosophical logic and, much more surprisingly, with the relevant literature in social psychology and judgement and decision making. We hope to take steps in this chapter towards a better integration of the research on counterfactuals.

## The philosophical analysis of counterfactuals

Analytical philosophers and philosophical logicians (going back to Chisolm, 1946; Goodman, 1947) have given sophisticated analyses of counterfactual conditionals and developed formal theories of them. We will focus on the philosophical work that has had the greatest influence on us.

We have already described, in Chapter 2, the T2 account of indicative conditionals in Stalnaker (1968). Here we will say more about Stalnaker's general approach and about how he extends it to counterfactuals. Stalnaker is not one of those philosophers who see a sharp distinction between indicatives and counterfactuals, and he applies his version of T2 to both. Compare the counterfactual 7.4—'if I had taken that pawn, I would have won the game'—with the indicative conditional:

7.5   If I take that pawn, then I will win the game

We might have had the thought expressed by 7.5 in the course of the game or asserted it during a break in the game. Some researchers have argued that counterfactuals such as 7.4 and indicatives such as 7.5 are closely related (Dudman, 1984), and Stalnaker (1968) applies essentially the same theory to both. Stalnaker's theory makes use of 'possible-world semantics', but a 'possible world', in his theory, is better termed 'an alternative way *the* world may be or might have been' (Stalnaker, 1975). Considering alternatives in this sense is necessary in the mental processes of probability judgement and making decision, and Stalnaker grounds what he says about these alternatives in those mental processes. The logically possible alternatives for both 'possible world' semantics and ordinary decision making in our chess example are by now well known to us in this book. These alternatives are the TT, TF, FT, and FF possible states of affairs:

(TT) I take the pawn and I win the game
(TF) I take the pawn and I do not win the game
(FT) I do not take the pawn and I win the game
(FF) I do not take the pawn and I do not win the game

These possible states would be relevant to our decision making in the game and the conditionals we would assert about it.

Possible states of affairs can be considered in more or less detail in our decision making. In the game, we would think about the possible options of taking the pawn or, alternatively, not taking the pawn. By reflecting on the probabilities and utilities of the possible outcomes, TT, TF, FT, and FF, we could make a decision about the best option to take. Assuming that the only utility for us comes from winning, we want to choose the option that makes it most likely we will win. In our example, the probability of TT is much greater than the probability of TF. With the other option, of not taking the pawn, let us suppose that the probability of FT is practically 0. The best move for us is to take the pawn. We might be able to work out in the game that this is our best move. The last step is grasping this fact may be having the conditional thought expressed by 7.5. We might also assert 7.5, perhaps to get our opponent to resign if the probability of TF is close to 0. The grounds for asserting the counterfactual 7.4, in retrospect, are essentially the same ones we would have had for asserting the indicative 7.5 if we had realized that taking the pawn was our best move at that point in the game.

As we explained in Chapter 2, Stalnaker (1968) modified the Ramsey test so that it could be applied to cases in which we must make hypothetical changes to our existing assumptions or beliefs. In Stalnaker's account, people would use the original Ramsey test to judge 7.5 and his modification of it to judge 7.4. Considering 7.5 first, we would begin the Ramsey test by hypothetically assuming the antecedent, that we take the pawn. We would then finish the test by inferring what degree of belief we could have in its consequent, that we will win the game. In this example, it is quite clear how we should make this inference. We should do this by using our understanding of chess to work out the possible moves that could follow our capturing the pawn. The result would be a conditional probability judgement at the point in the game where we were trying to decide whether to capture the pawn.

Now suppose we have lost the game by not taking the pawn, and are considering 7.4 as we review the game later. We cannot, at this point, use the test as originally described by Ramsey (1931/1990), which presupposes that we do not know that the antecedent of 7.4 is false. Stalnaker (1968) extended the Ramsey test to deal with this kind of case. Using this extension, we are to assume the antecedent of 7.4 while making minimal hypothetical changes to our beliefs to preserve consistency. The changes are to be as small as we can reasonably make them. It would be unjustified to suppose that the rules of chess, or still less the laws of nature, would change merely because we captured the pawn rather than made some other move. More than this, it would be grossly irrelevant. We are going over the game again to see what we can learn to improve our decision making in future games. This will be a pointless exercise if we assume, in our hypothetical thought about the game, that its rules are different from what they actually are.

In Chapter 8, we discuss in detail experimentation that shows that in general people tend to judge the probability of the indicative conditional 'if p then q' as being equal to that of the conditional probability, $P(q|p)$. We find this both for abstract conditionals with defined frequency distributions and for thematic causal conditionals where the relevant probabilities are supplied by prior belief. These findings clearly support the hypothesis that people evaluate such conditionals by use of the Ramsey test. The result of applying Stalnaker's extension of the Ramsey test to counterfactual conditionals should be to generate similar conditional probabilities. If we believe that the indicative conditional 7.5 has a high probability, then we should assign the same probability to the counterfactual 7.4. We report a recent experiment of our own in Chapter 8 that supports this hypothesis.

## Possible ways the world may be or might have been

By making minimal changes to our beliefs, Stalnaker holds that we are thinking about an alternative way the world may be, might have been, that is minimally different from the way it actually turned out. Another, more metaphorical way to express this is that we are thinking about the 'closest' or 'nearest' possible way the world may be or might have been. The metaphor is appropriate as people often do have firm intuitions about which possibilities are closer or nearer than others to the way things actually are (Kahneman and Miller, 1986; Kahneman and Varey, 1990; Medvec *et al.*, 1995; Roese and Olson, 1995; Roese, 1997, 2004). For our example of 7.4, we are to consider the closest possibility to what actually happened in which we take the pawn. As we have already pointed out, we might find this intuitively a really close possibility. We might even say, with great regret, that we were 'very close' to winning the game or 'nearly' won it. We 'snatched defeat from the jaws of victory' by hesitating when we would normally capture a piece.

People will sometimes suffer from biases in their assessments of closeness (Teigen, 1998, 2004; Tetlock, 2002). We may like to imagine ourselves to be bold chess players when we are really timid. The rare occasions when we were bold in our play may be more available in our memories than the more numerous times when we were timid.

The availability heuristic (Tversky and Kahneman, 1973; Shwarz and Vaughn, 2002) may then lead us to judge that we were close to taking the pawn or nearly took it. We might also associate a high conditional probability with 7.4 because the occasions on which we were bold and won are more available to us than the occasions on which we were bold and lost. We might tend to forget the times when we were bold in taking a piece but then lost our nerve and the game by becoming timid again. Being nervous in the heat of the actual game, we might associate a low conditional probability with 7.5, but then later in reviewing the game associate, thanks to the availability heuristic, a high conditional probability with 7.4.

Kahneman and Tversky (1982) also proposed what they called 'the simulation heuristic', which can be used to make conditional probability judgements and assess counterfactuals. The idea is that a 'simulation model' can be 'run' from a possible condition or what will be the antecedent of a counterfactual. There are clear similarities between this heuristic and the Ramsey test, the original version and Stalnaker's extension of it. In our example, there is a clear sense in which a simulation could be run of the possible game on the supposition that we took the pawn. Indeed this simulation could be done on a computer. We are attracted to this account, but are cautious about the use of the term 'heuristic' to describe it. In our previous discussions of dual process theory, we have generally associated the term 'heuristic' with automatic and rapid processes in System 1. However, the process that Kahneman and Tversky described in this paper would definitely engage some explicit reasoning in System 2, in order to conduct the mental simulation. For example, evaluating alternative continuations in a chess game, requires an element of explicit calculation. In the more general decision-making context to which Kahneman and Tversky refer, simulations would typically involve explicit reasoning also, even though the failure to consider alternative pathways that they describe might be a System 1 effect, analogous to some of the relevance effects.

Kahneman and Tversky (1982) describe biases that we could suffer from when running a simulation, and these could equally affect a Ramsey test (see the role of System 1 and System 2 thinking in the Ramsey test discussed in Chapter 9). There are of course the biases associated with the availability heuristic. More subtle bias could result from insufficient attention to relatively improbable events in an extended simulation. We might truly be relatively bold players, and knowing this, we might reject any possible timid moves as part of our simulation. Yet the very fact that we did not take the pawn in the first place shows that we are sometimes timid. We should not completely rule out some timid play on our part in an extended simulation, but we still might tend to do this.

All this is not to say that simulation, or the Ramsey test, will usually be badly biased. Tversky and Kahneman (1973) themselves pointed out that availability can be an 'ecologically valid cue' for a frequency judgement. We would be less prone to biases if we relied on written or, even more, computer records of our chess games to help System 2 counteract any System 1 biases. This could give us good justification for judgements about our style of play and how close we were to taking the pawn and winning. Applications of the Ramsey test can be grounded in objective frequencies and be much more than speculation about supposedly fanciful counterfactual possibilities. A rigorous computer simulation of our taking the pawn could result in a win for us in almost every outcome. That would show that we were well justified in asserting 7.4 and could learn something from it to help us win matches in the future.

A significant virtue of Stalnaker's view of a possibility as 'an alternative way the world may be or might have been' is the way it depends on what is relevant in a particular context of decision making and communication. Some possible alternatives are relevant in some contexts but not others, just as some of our beliefs, but not others, are relevant to some inferences and decisions, but not others. Consider these famous examples (Adams, 1970) from philosophical logic:

7.6   If Oswald didn't kill Kennedy, then someone else did
7.7   If Oswald hadn't killed Kennedy, then some else would have

These examples are usually used to show that counterfactuals can be quite different from some indicative conditionals, however closely related they may be to other indicative conditionals (Edgington, 1995, 1997; Woods, 1997). Most people think that 7.6 is true and 7.7 is false.

A counterfactual such as 7.4 may be related to an earlier use of an indicative such as 7.5. But in the following, the past tense indicative 7.8 is definitely different from the past tense counterfactual 7.9.

7.8   If I took that pawn, then I made a real blunder later
7.9   If I had taken the pawn, then I would have made a real blunder later

A 'blunder' is chess jargon for an erroneous move out of relation to the general playing ability of the chess player. We think that some important issues are clearer in our example than in the one about Oswald and Kennedy, and so we will discuss our example in more detail.

Suppose that, while reviewing the written record of our game, we recall quite clearly that we lost the game. Now we come to the bottom of the page in our booklet and see that taking the pawn was a possible move. We cannot definitely remember whether we took the pawn or not. In this context, we will consider 7.8 to be probably true and 7.9 probably false. It is clear that both judgements can be given an objective grounding, but in different ways. Computer simulations could best illustrate this. The simulation for 7.9 would begin with taking the pawn in the given position. We have assumed that in most of the possible outcomes from this point, according to chess theory, the result is a win for us. The consequent of 7.9 is false in most of these possible outcomes, which are all ones in which the antecedent of 7.9 is true, and thus 7.9 is probably false. (Of course, we might also have to factor in our tendency to blunder under the stress of match play.) In contrast, the computer simulation of 7.8 would operate under the supposition that *we end up losing the game*. From taking the pawn under this condition, which ought to be a winning move, most of the possible outcomes will be ones in which we made stupid mistakes, and so 7.8 is probably true. The difference between 7.8 and 7.9 is that for the evaluation of 7.8, but not for 7.9, the relevant possibilities are only ones in which we lose the game.

We conceive of the simulation 'heuristic' as one way in which the Ramsey test can be implemented. Both the heuristic and the test have so far only been sketchily described and should be studied much more experimentally. But both have a great deal of psychological plausibility. In our view, the Ramsey test can be developed as a psychological hypothesis by studying in detail how it is implemented by System 1 heuristics and System 2 reasoning. But as we explained in Chapter 2, there is a problem for

Stalnaker if the Ramsey test of a conditional always yields the conditional probability of the consequent given the antecedent. The probability of the Stalnaker conditional, whether this is expressed as an indicative or as a counterfactual, cannot always be identified with the conditional probability. As we explained in Chapter 2, this follows from the proof in Lewis (1976). We have also already pointed out that, in light of this proof, there is prima facie evidence against the hypothesis that the indicative conditional of natural language is the Stalnaker conditional. In experiments run so far, people have judged the probability of certain indicative conditionals to be the conditional probability. However, further experiments are definitely called for on probability judgements about ordinary conditionals. Some of these experiments should be on counterfactuals such as 7.4 and their relation to indicative conditionals such as 7.5, and on the relation between degrees of confidence in both of these conditional types and conditional probability judgements.

Psychologists who study reasoning have not so far published any studies of people's probability judgements about counterfactuals. The single experiment that we have recently run, reported in Chapter 8, is the only one that we are aware of on this topic. They have also done very little work on what people judge to be logically valid inferences for counterfactuals. Take the simple question, 'Does a counterfactual such as 7.4 logically imply that its antecedent is false?' Most philosophical logicians, including Stalnaker, would answer 'No' to this question. They would claim instead that a counterfactual only pragmatically suggests that its antecedent is false. The intuitive argument for this pragmatic position is strong. Suppose we are reviewing an old game and do not remember it well. We are under the mistaken impression that we did not take the pawn, when actually we did. With this mistaken belief, we will naturally assert a counterfactual such as 7.4 at the place in the written record of the game where taking the pawn was an option. If we turn the page and find we did take the pawn, and went on to win the game, we surely will not accept that we stated something false in asserting 7.4, but rather claim that the content of what we asserted was true.

Social psychologists and psychologists working on judgement and decision making have done the most extensive research on counterfactuals. But psychologists of reasoning have only recently started to investigate, in any depth, inferences to do with counterfactuals. Thompson and Byrne (2002) confirmed that people do infer that the antecedent of a conditional such as 7.4 is false, but this work does not yet discriminate between pragmatic inferences and logical inferences. Thompson and Byrne asked their participants what a speaker 'meant to imply' by asserting a counterfactual such as 7.4. In our view, this kind of question encourages pragmatic inferences. Our hypothesis is that people will not infer that 7.4 is actually false given that its antecedent and consequent are true, but this hypothesis has yet to be confirmed by experiments.

Psychologists of reasoning are even farther from discriminating between some of the different logical systems for the counterfactual. Lewis (1973) criticized aspects of Stalnaker's system. In particular, Lewis rejected Stalnaker's position that there is exactly one possible alternative way that the world might have been in which the antecedent of a counterfactual is true. When we use Stalnaker's generalization of the Ramsey test on 7.4, we consider a number of ways the game might have developed given that we took the pawn. This mental process will give us a degree of belief in the consequent of 7.4 given its antecedent. But according to Stalnaker, we are trying in this process to represent accurately just one possibility: the closest in which we take the pawn. In Stalnaker's system, 7.4 is true if and only if the consequent of 7.4 is true in that single, closest possibility. Lewis doubted that there is always such a unique closest possibility, and even held that there might be closer and closer possibilities, without end, in which the antecedent of a counterfactual is true. The result is that the logic of what we can call the Lewis conditional differs in certain respects from that of the Stalnaker conditional. Lewis (1973, p. 80) goes so far as to admit that an 'ordinary language speaker' might agree more with Stalnaker, but psychologists of reasoning have not run experiments to test whether this is so.

Psychologist of reasoning have may have been slow to run experiments on different 'possible world' analyses of conditionals because they have reacted so negatively to Lewis's metaphysics. Lewis (1973; 1986) was an extreme realist about possible worlds: he held that other possible worlds are just like the actual world. We rightly call the world we live in 'the actual world', but according to Lewis, people in other possible worlds rightly call their worlds 'the actual world'. There is nothing intrinsically special about what is called 'the actual world' by whoever is doing the calling. Lewis accepted that this realism has counterintuitive results that go strongly against common sense (see especially Lewis, 1986, pp. 133–135). One of these is that we cannot be present in more than one possible world. We cannot be present in a possible world in which we take the pawn and win the chess match, as we would then absurdly be in more than one 'actual' world. Thus Lewis claims that 7.4 is true if and only if a 'counterpart' of us (someone very like us) wins the game in the closest possible world (or in a series of ever closer possible worlds) in which a 'counterpart' takes the pawn. Lewis has much to say (in especially Lewis, 1986) to try to clarify his notion of a possible world (and of similarity between possible worlds as he sees them) and of 'counterparts'.

We certainly reject Lewis's type of realism about possibilities, and to stress this, we avoid the term 'possible world' as much as we can. It is psychologically bizarre to claim that people are thinking about a 'counterpart' of themselves when they reflect on actions they might have taken but did not, or feel relief or regret about the way things might have been. However, we will not argue a case against Lewis (see Stalnaker, 1984), nor argue for any particular stand on the metaphysics of possibilities. We agree with Stalnaker (1975, 1976) that ways *the* world may be, or might have been, necessarily arise in probability judgement and decision making (see also Kripke, 1980). These 'possible outcomes' or 'possible states of affairs' are the very basis of hypothetical thought. Reference is made to them in so-called 'possible world' analyses of conditionals, but that is equally true in theories of probability judgement and decision making. Referring to these possibilities in the psychological study of judgement and decision making, or of emotions such as relief or regret, is not to take any particular metaphysical view of them. This is equally true of referring to them in psychological studies of indicative and counterfactual conditionals. (See Lowe, 2002, chapter 7, for an especially clear introduction to the metaphysics of 'possible worlds' and to conditionals and metaphysics.)

## Counterfactuals and the psychology of reasoning

We have pointed out in Chapter 4 that Rips and Marcus (1977) have the psychological T2 account of ordinary conditionals that is closest to Stalnaker's analysis in philosophical logic. In fact, their account is in some ways more like a 'possible world' analysis in logical semantics than a psychological hypothesis. As we also argue in Chapter 4, their account is unbounded, and we will briefly illustrate that again here with our example of a counterfactual, 7.4.

Rips and Marcus would tell us to take the antecedent of 7.4, that we took the pawn, as a 'seed' supposition for evaluating 7.4. We are to rank our current beliefs from those we can most easily give up hypothetically, at the highest level, to those that we would be least inclined to suspend even hypothetically, at the lowest level. At the highest level will be our belief that we did not, in fact, take the pawn. At lower levels will be our beliefs about the rules of chess and the physiological and physical facts that would enable us to move pieces around the board. We add beliefs at the highest level that are consistent with the suppositional 'seed' and carry on in this way through the levels. We will often have to divide the consistent set we are creating in this way into two or more consistent sets, each of which we will then have to expand as we go through the ranks. Supposing that we captured the pawn, it will then be our opponents move. He will be able to make a certain number of moves, by the rules, in reply, and we should, by what Rips and Marcus say, start to build consistent sets for each of these possible counter-moves. Obviously, this procedure is an impossible one for us to carry out mentally even at this early point. We could theoretically program a (very big) computer to do it, and hold that 7.4 is probably true if and if only if we win the game in most of the consistent sets running the program would create. But this definition is much more like a logical analysis of what it means to state that 7.4 is probably true than a plausible psychological account of why we might think that 7.4 was probably true.

A much more psychological account would make use of the simulation heuristic. The idea would be that we mentally simulated the game from the supposition that we took the pawn. In this simulation, we might consider only a small number of the possible countermoves our opponent could make, the most reasonable ones, and a similarly small number of further moves by ourselves and so on. In the case of chess, calculation of continuations by human players must be sharply limited in this way to avoid combinatorial explosion of the number of possibilities to be considered. Chess-playing computer programs, using brute-force methods, compute much larger numbers of continuations than any human player could or does. Hence, the intelligence and quality of human play depends critically on how effective the heuristic elimination of most possible moves actually is.

Our mental simulation (assuming we are expert chess players) would depend on some limited calculation tied to holistic evaluation of the resulting possible positions. If these are favourable then we will again come to the conclusion that 7.4 is probably true. An even more plausible psychological proposal would be that we would do little or nothing in the way of calculation of continuations, but rather make a judgement directly about the quality of the position taking the pawn would have put us in. Our experience of past games might enable us to recognize this position as a strong one, and on that basis alone we might conclude 7.4 is probably true. To take a different example, if we have a choice of a game of golf or cricket at the weekend, we might simply decide to play cricket because we generally prefer that game. Or we might do rather more cognitive work, taking into account the weather forecast, the quality of the opposition and the company and so on, which might by mental simulation of the two choices lead us to believe that on this particular occasion the golf would be more enjoyable. There are other possibilities for how people might make these kinds of judgement, and no doubt there would be some individual differences that could be investigated experimentally. But we can be sure, given what we already know about people's working memory limitations, that no one would get very far in constructing great consistent sets in the way described by Rips and Marcus.

We also pointed out the problems in Chapter 4 of the psychological account of conditionals in Braine and O'Brien (1991). Basically, they confuse a procedure for settling the logical question of whether conditional conclusions validly follow from premises with the empirical question whether there is good justification for asserting or believing the conditionals. Consider what they say about the following instance of strengthening the antecedent:

7.10   If this match were struck, it would light
7.11   If this match were soaked overnight and were struck, it would light

Inferring 7.11 from 7.10 is logically valid if these are material conditionals, but is invalid if these are Stalnaker or Lewis conditionals. Braine and O'Brien claim (p. 190) that this inference is sometimes valid and sometimes invalid. The inference is invalid if the match is a standard one that will not light when wet, but is valid if the match is a special make that will light when wet.

It is a fundamental misunderstanding of validity to imagine that the validity of an inference can depend on a matter of fact, such as whether a match is of some special make. Validity is determined by logical form, not matters of fact. We could be justified in asserting and believing both 7.10 and 7.11. Using the Ramsey test, and supposing that the match is a special make that will light when wet, we can infer high degrees of belief in the consequents of 7.10 and 7.11, giving us high confidence in 7.10 and 7.11 at the same time. But this fact does not mean that 7.11 follows logically from 7.10. It is essential to distinguish what logically follows from what justifiably follows in a wider sense when applying the Ramsey test. The test can be used to make logical inferences, but it can also be applied, much more widely, to give us more or less confidence in conditionals because of matters of fact that we are aware of. Braine and O'Brien's theory, as it stands, cannot even tell us how a counterfactual can have different degrees of probability.

Johnson-Laird and Byrne (1991, 2002) have a mental model psychological theory of counterfactuals that is more interesting, in our view, than the mental logic account of Braine and O'Brien. Johnson-Laird and Byrne's mental models for 7.4 would be:

| Fact: | I did not take the pawn | I did not win |
|---|---|---|
| Counterfactual possibilities: | I took the pawn | I won |
| | I did not take the pawn | I won |

The above mental models begin by representing the actual state of affairs in which we did not take the pawn and did not win. The mental models then represent two counterfactual possibilities. The first is one in which we took the pawn and won, and the second is one in which we did not take the pawn and won.

The first problem with these mental models is that they apparently represent 7.4 as logically implying that its antecedent and consequent are actually false. As we have argued, a counterfactual is not false when its antecedent and consequent turn out, in fact, to be true. Nor is a counterfactual probably false whenever its antecedent and consequent are likely to be true. Someone might assert 7.4 because he has a poor memory, forgetting that he did take the pawn and won, or when he is unsure whether or not he took the pawn and won. Discovering that his memory is faulty, and that he did take the pawn and win, he can rightly conclude that 7.4 is true, not false. After all, what better evidence could he have for 7.4? The grammatical form of 7.4 may have been misleading in the context, but that does not make 7.4 false.

The second problem with the mental models is that they have too little representational content. Consider how the above mental models represent the counterfactual possibility that we took the pawn and won the game. This representation merely asserts, in effect, that such a counterfactual possibility exists. It does *not* represent whether that possibility is a very close one or a very distant one. That being so, the mental models cannot explain why we would feel perhaps bitter regret when we nearly took the pawn, but changed our mind for no good reason, and then realized just too late what a decisive move it would have been.

The above mental models also cannot explain why we would assert or believe 7.4 in some cases but not others, depending on our subjective judgements. Whether we saw taking the pawn as a good move or a bad move, 7.4 would have exactly the same mental models. Suppose taking the pawn was such a bad move that we would never contemplate it. Even in that case, there is a counterfactual possibility in which we took the pawn and won. That is a counterfactual possibility in which our opponent resigns when he is in an obvious winning position. Even if he is a really astute player, the counterfactual possibility exists in which he resigns in such a case. We would judge this resignation to be a highly unlikely and distant possibility, but Johnson-Laird and Byrne's mental models do not have the structure to represent subjective probability or closeness judgements about counterfactual possibilities. This limitation is serious. We would only assert or believe 7.4 if we thought that our taking the pawn and winning was a close possibility. We would not bother with 7.4 if winning after taking the pawn depended on our opponent making a highly uncharacteristic blunder. That counterfactual possibility would be too distant for us to learn something from for our future decision making. Johnson-Laird and Byrne's mental models cannot explain such basic facts about counterfactual assertion and belief and the decision making that is so closely related to it.

We can sharpen our case against the mental models by comparing 7.4—if I had taken that pawn, then I would have won the game—with the following:

7.12   If I had taken that pawn, then there would have been stalemate

This counterfactual has these Johnson-Laird and Byrne mental models (compare with those given above for 7.4):

| Fact: | I did not take the pawn | There was not stalemate |
|---|---|---|
| Counterfactual | I took the pawn | There was stalemate |
| possibilities: | I did not take the pawn | There was stalemate |

We would not assert or believe 7.4 and 7.12 at the same time, and yet Johnson-Laird and Byrne's mental models cannot explain this fact. There is the counterfactual possibility in which we take the pawn and win, and there is the counterfactual possibility in which we take the pawn and there is stalemate. Both of these possibilities exist, and both are equally represented in the Johnson-Laird and Byrne model models. These models do not represent one of these possibilities as closer or more plausible than the other. But suppose we can see that there could be a stalemate after we took the pawn only if we made highly uncharacteristic blunders. This judgement of ours is not reflected in any aspect of the Johnson-Laird and Byrne mental models, and so these models cannot explain why we would assert and believe 7.4 in this case but not 7.12 (see Over, 2004a,b, on problems with Johnson-Laird and Byrne's mental models of counterfactuals).

Johnson-Laird and Byrne (2002, p. 652) make the comment about counterfactual possibilities: 'One cannot observe counterfactual states, and so the truth or falsity of a counterfactual conditional about a contingent matter may never be ascertained. Perhaps that is why counterfactual speculations so intrigue historians, novelists, and sports fans.'

Our example 7.4 is a counterfactual statement that might be made by a chess fan, and we have seen that its probable truth can be inferred from heuristics that, sometimes at least, are well grounded in objective facts. Thinking about 7.4 is not necessarily idle speculation, but can be of real help in future decision making. Johnson-Laird and Byrne may be imagining that a counterfactual possibility has to be a Lewis-type 'possible world'. It is indeed a possible criticism of Lewis that, if possibilities were as he described them, then we could have no knowledge of them. However, this criticism does not apply to Stalnaker's philosophical analysis of counterfactual possibilities. Still less can the criticism be levelled at the pre-theoretical concept of a counterfactual possibility that we all rely on when we asserted or believe counterfactuals or use them in decision making.

It might be argued that we have chosen our chess example carefully and that not all counterfactuals can be so well justified. However, there are many perfectly ordinary counterfactuals that can be well justified. Consider:

7.13   If we were to eat the chicken then we would get sick

We might assert 7.13 about some only partially cooked chicken that has been left out in the sun at a picnic. This assertion would seem to be well justified, and if more justification is called for, we could have the chicken tested for salmonella bacteria. It is not useless speculation to think about 7.13. Our resulting confidence in 7.13 might be the reason that we decide not to eat the chicken. When the lab report comes back positive, or other people who did eat the chicken get sick, we will rightly feel relief that we did think about 7.13 and the closeness of the counterfactual possibility that it represents.

## Counterfactuals and causation

The view that counterfactuals and causation are related to each other goes some way back in philosophy (Hume, 1902, section 60; Lewis, 1973). Return to the example we have just given. If eating the picnic chicken causes sickness, then 7.13 clearly holds on any reasonable account. But can 7.13 be true when eating the chicken at the picnic does not cause sickness? It may be the water that everyone drinks at the picnic grounds that causes the sickness. However, that possibility is ruled out if the following is true as well as 7.13:

7.14   If we were not to eat the chicken then we would not get sick

Eating the chicken would make us sick, and not some other factor or condition, if both 7.13 and 7.14 hold.

Let us consider how we would get evidence about a causal relation. Suppose that a number of people at the picnic have eaten the chicken and a number have got sick. Eating the chicken may be the cause of the sickness, but so might drinking the water at the picnic grounds or some other factor. We can make progress by trying to establish the degree of covariation between eating the chicken and becoming sick. Yet again we have four possible states of affairs:

(TT) Eating the chicken and becoming sick
(TF) Eating the chicken and not becoming sick
(FT) Not eating the chicken and becoming sick
(FF) Not eating the chicken and not becoming sick

Studying people at the picnic will tell us how many of them fall under each of these possibilities. For summarizing data in cases like this, we can use 2 × 2 contingency tables, with four cells conventionally labelled a, b, c, and d. The number of TT cases (of people who ate the chicken and got sick) goes into the a cell. The number of TF cases (of chicken and no sickness) into the b cell, the number of FT cases (of no chicken and sickness) into the c cell, and the number of FF cases (of no chicken and no sickness) into the d cell. When we have sampled a number of people at the picnic in each of these four cells, we can work out the $\Delta$p statistic (Allan, 1980; Shanks, 1995, 2004) in this way:

$$a/(a + b) - c/(c + d)$$

The $\Delta$p statistic represents the difference between two conditional probabilities, $P(q|p) - P(q|\neg p)$. When $P(q|p) = P(q|\neg p)$, p and q are independent and p cannot cause q. When $P(q|p) > P(q|\neg p)$, this 'delta p rule' measures the extent to which p *raises* the probability of q.

If the above difference is 0, this is evidence that eating the chicken is independent of becoming sick and so that there is no causal structure connecting the two. But suppose the difference is greater than 0. This is evidence of some kind of causal structure connecting eating the chicken and getting sick. Eating the chicken may not directly cause the sickness. There might be some underlying cause that connects both, such as a virus that makes people hungry and then makes them sick. But this possibility is not plausible in this case, and we could reasonably infer that eating the chicken does cause the sickness.

There are extensive psychological studies of how people use covariation information, which can be explicitly given to them in 2 × 2 tables, to make inferences about causation. The extent to which they are rational in their responses has been subject to considerable debate (Anderson, 1990; Shanks, 1995; Stanovich and West, 1998b, 2003; Over and Green, 2001). However, people clearly would not speak explicitly in terms of covariation, 2 × 2 tables, and the $\Delta$p statistic if they were trying to decide whether to eat the chicken in an actual case, such as our picnic. They would instead think about conditionals such as 7.13 and 7.14, and make a decision based on their confidence in these.

The point we wish to make here is that people could, in effect, make proper use of the $\Delta$p statistic by applying the Ramsey test to 7.13 and 7.14. Making the hypothetical supposition that the antecedent of 7.13 holds, our no doubt highly available knowledge, of people who ate the chicken and got sick, would give us a high degree of confidence that the consequent of 7.13 follows. In other words, we would make the judgement that the conditional probability is high of our becoming sick given that we eat the chicken, P(sickness/chicken). This first use of the Ramsey test would give us high confidence in 7.13. Then making the hypothetical supposition that the antecedent of 7.14 holds, our relevant available knowledge, of people who did not eat the chicken and did not get sick, would make us highly confident that the consequent of 7.13 follows. Our judgement would be that the conditional probability is high, of our not becoming sick, given that we did not eat the chicken, P(not-sickness/not-chicken). It would follow that we had high confidence in 7.14. With our confidence high in both 7.13 and 7.14, we would decide not to eat the chicken.

Recall that when P(not-sickness/not-chicken) is high, P(sickness/not-chicken) will be low, and suppose that our judgement is coherent in this respect. Then with high confidence in both 7.13 and 7.14, via the Ramsey test, we have implicitly made the right use of the equivalent of the $\Delta p$ statistic here:

$$P(\text{sickness/chicken}) - P(\text{sickness/not-chicken})$$

The above difference will be implicitly high for us, and we have, in effect, inferred the right causal conclusion by having high confidence in both 7.13 and 7.14. We have reached this conclusion by applying the Ramsey test to 7.13 and 7.14.

Sometimes available covariation information could be used too naively in a Ramsey test. For example, we may remember that it has rained many times when we have gone out with an umbrella, and rained few times when we have gone out without an umbrella. The difference given by the $\Delta p$ statistic is high in this example, but obviously it would be wrong to conclude that our going out with an umbrella *causes* the rain. There is an underlying causal structure that accounts for the association between the two events. A weather system causes the rain and causes a weather forecast to be made, and our hearing the weather report causes us to go out with the umbrella.

The correct thing to say about causation in this weather example is clear, but consider the following conditionals about a day when sunny weather is correctly predicted:

7.15 If I were to go out today with an umbrella, then it would rain
7.16 If were not to go out today with an umbrella, then it would not rain

A naive use of the availability heuristic in Ramsey tests on 7.15 and 7.16 would make it appear that P(rain/umbrella) and P(not-rain/not-umbrella) are both high. That would give us high confidence in 7.15 and 7.16, and we might well decide on this basic to take our umbrella when we went out. This would be a biased judgement and a bad decision. Of course, people are sometimes biased in their judgements and decision making, and they will, in particular, sometimes engage in 'magical' thought and misrepresent the true causal structure of the world. Perhaps there are even gardeners who take umbrellas outside when they are desperate for rain. But people are not always as irrational as that, and they can make correct judgements in examples like this one. The question is how they do it.

People might restrict, in Ramsey tests on 7.15 and 7.16, the past instances they can recall to times when they took an umbrella out after hearing a weather forecast of sunny weather. If so, they will judge P(rain/umbrella) to be low and avoid a bias. But suppose that there are no past cases when we have gone out with an umbrella after hearing a forecast of sunny weather. Why have we always decided to leave our umbrella behind in these cases, and why will we make this decision again now? One promising suggestion is that we make use of a mental model of the relevant causal relations. We may represent the causal structure, with the underlying cause as weather systems, which determine both the weather and forecasts about it, followed by our taking the umbrella out. Such a mental model must, of course, have more representational structure in it than the Johnson-Laird and Byrne mental models of 7.15 and 7.16.

Pearl (2000) has presented a normative theory in which causal structures are represented in causal diagrams or graphs. He proposes (p. 239) that, to judge a counterfactual such as 7.15, we perform a 'minisurgery' on our causal model, which is the minimal change necessary in the model to make the antecedent true. We think that this process could be seen as another way of implementing the Ramsey test, or at least a version or extension of it. Consider our use of hypothetical reasoning in an example in which we know that sunny weather is forecast. We hypothetically suppose the antecedent of 7.15: that we go out with our umbrella. What confidence should we have in the consequent that it will rain under this supposition? Our causal model of what causes both the weather and weather forecasts is relevant. But there has been a forecast of sunny weather and not of rain. The model will tell us that a weather system is on the way that will bring sunny weather. It would not be a minimal change to this model to make it represent rainy weather in this example, merely because we go out with an umbrella. We therefore have very low confidence in 7.15. More hypothetical thought using the causal model would give us, for essentially the same reasons, very high confidence in 7.16. The final result would be that we decided not to take our umbrella out with us. This kind of hypothetical thought, which is grounded in a causal model, could be interpreted as yet another form of the Ramsey test.

The result of using a causal model in such a Ramsey test is that our confidence in 7.15 is very low. But we also saw that we could get very high confidence in 7.15 in another implementation of the Ramsey test by reflecting only on the past relative frequency of rain, given our going out with an umbrella. This naive use of the relative frequency would give us a high value for P(rain/umbrella) and lead to a bad decision to take the umbrella out, as we also saw. Using the causal model in a Ramsey test would give a different value to 7.15 than using the past relative frequency in a Ramsey test to get the conditional probability, P(rain/umbrella), as the probability of 7.15. Basing the test on the causal model would result in an unbiased judgement about 7.15 and better decision making.

We should again remember our explanation in Chapter 2 of how Lewis (1976) showed that the probability of a Stalnaker conditional is not generally identical with the corresponding conditional probability. In fact, Lewis's result applies to any system for ordinary conditionals, including his own, in which the conditionals are true or false at each possible state of affairs. At the same time, there are arguments that standard decision theory sometimes justifies bad decisions because it is based on conditional probability judgements rather than judgements about causal structures. It might, for example, justify 'magical' thinking in gardeners who want rain and try to get it by using their umbrellas. To avoid such problems, the critics have argued that standard decision theory should be replaced by causal decision theory. In causal decision theory, decisions are based on the probabilities of Stalnaker or Lewis type counterfactuals, which can sometimes better reflect causal structure than conditional probabilities, according to the critics (Gibbard and Harper, 1978; Lewis, 1981; Joyce, 1999; Pearl, 2000).

Do ordinary people ever evaluate the probabilities of counterfactuals, and possibly related indicative conditionals, by using causal models? Do they then ever assign these conditionals a probability that is not equal to a corresponding conditional probability? Do people make decisions in line with standard decision theory or causal decision theory when these differ in their recommendations? Our position is that much more psychological theorizing must be done, and experimental studies run, before psychologists can answer these questions. We must understand much more deeply how people engage in hypothetical thought, implementing in effect different versions of the Ramsey test, and make probability judgements on this basis. With this understanding, psychological research could greatly increase its relevance to debates about the best logical system for ordinary conditionals and the best normative theory for decision making. People do not always display biased counterfactual judgement and reasoning. Sometimes they employ counterfactuals to good effect in their reasoning and decision making. Finding out how they accomplish this in a bounded but efficient way could well suggest normative standards for how this should be done.

## Counterfactuals and conditional probability

Lewis's proof, that the probability of a T2 type conditional does not in general equal the conditional probability, assumed that the conditional was true or false at every possibility. There are at least some reasons for holding that this assumption is true of counterfactuals, if not for all indicatives. It follows that there are some reasons for preferring T2, or a Stalnaker type analysis, for counterfactuals.

Many counterfactuals are closely related, if not logically equivalent, to statements about dispositions and capacities. For example, consider again the example in Chapter 2 of our bone china Spode teapot. We take great care not to drop it on the flagstone floor in our kitchen. Why is that? The reason is obvious: the teapot is extremely fragile. In Chapter 3, we reviewed experiments in which participants are asked to evaluate the truth value of indicative conditionals about frequency distributions, e.g. 'If there is a B written on the left side of the card then there is a 5 written on the right.' The participants tend to judge that a false antecedent case, e.g. the card drawn has an E on the left, is 'irrelevant' to whether the indicative conditional is true or false. We might interpret these participants as claiming that the conditional is neither true nor false in the false antecedent case. But suppose we tell the participants in an experiment that we are not, in fact, going to drop our extremely fragile teapot on our stone kitchen floor. They will surely not respond by saying that this information is 'irrelevant' to the truth or falsity of the counterfactual, 'the teapot would break if we were to drop it', and may well claim that it means this counterfactual is true.

The false antecedent information for the counterfactual is much more revealing than the false antecedent information for the indicative conditional about letters and numbers on cards. The point is that there is so much more false antecedent information to give in the teapot example. In the card example, we have no more information about the false antecedent case than that it is one of drawing a card with E on the left, which is indeed irrelevant. In the counterfactual example, we have significant information in the false antecedent case: it is a case in which an extremely fragile teapot is not dropped on a stone floor. That information does not appear irrelevant to the evaluation of the counterfactual.

Stalnaker's approach to the counterfactual example is natural and intuitive. The closest possibility in which we drop the fragile teapot on the stone floor is one in which the teapot breaks, and hence the counterfactual is true. Expressing the matter in terms of mental models, we would judge that the mental model in which the teapot is represented as dropped and breaking is 'closer' than one in which it is represented as dropped and not breaking. The former mental model is 'closer' than the latter because the former is minimally different from what we believe about the actual state of affairs, in which we know that the teapot is extremely fragile. We would consequently have high confidence in the counterfactual.

Lewis's proof shows that the probability of counterfactuals cannot in general equal the conditional probability, but the two probabilities could be very close in our example (see also Chapter 2). There is a relationship between people's judgements about the closeness of possibilities and their judgements about the probabilities of those possibilities (Teigen, 1998, 2004; Roese, 2004). Participants in an experiment would presumably judge the possibility in which the teapot drops and breaks as much closer to the actual state of affairs than the possibility in which the teapot drops and does not break. With this view of closeness, they would, explicitly or implicitly, assign a high conditional probability to the porcelain teapot breaking given that it drops. That means that we, as experimenters, could be hard put to tell whether there was a difference between their high confidence in the counterfactual and their high conditional probability judgement. If we changed the example to a non-fragile pewter teapot, people would have low confidence in the counterfactual, that the pewter teapot would break if it were dropped, and make a low conditional probability judgement that the pewter teapot breaks given that it drops. Again, it might be hard to discover whether there was a difference between these probability judgements.

We mentioned this problem in Chapter 2. We can test the hypothesis that an ordinary conditional, whether indicative or counterfactual, is a Stalnaker conditional, or a T2 conditional more generally, by seeing whether its judged probability equals the judged conditional probability. If the two probabilities are equal, that is evidence against the hypothesis, by Lewis's proof. If the two probabilities are not equal, that is evidence for the hypothesis, by Lewis's proof. But this proof does not imply that the two probabilities are far apart in experiments that are easy to design and run. It is much easier to think of experiments in which the probability of the truth functional conditional and the conditional probability are far apart. Some indicative conditionals at least may not express epistemic states, but may be very similar to, if not essentially the same as, counterfactuals in being about causal or other objective relations. Perhaps there should be a version of T2 for these indicative conditionals, as for counterfactuals. As we will describe in Chapter 8, we have found in experiments that the judged probability of causal indicative conditionals and counterfactuals is at least close to the conditional probability. In Chapter 9, we will discuss further how to try to decide between T2 or T3 for indicative conditionals.

# Conclusions

There may be a significant difference between indicative conditionals, or at least some of them, and counterfactuals. Some indicative conditionals at least, e.g. 7.2, 7.6, or 7.8, may express epistemic mental states, and counterfactuals may concern objective relations. Some philosophers have consequently argued for T3 for indicatives and T2 for counterfactuals (see Bennett, 2003, for an extended argument to this effect). However, some indicative conditionals, e.g. 7.5, might be closely related to counterfactuals and about causal or other objective relations, and that could imply T2 for them at least. Sadly, research on the psychology of counterfactual reasoning is not yet able to throw much light on these issues. The problem is that this research is poorly integrated, not only with philosophical work on counterfactuals, but even with the literature on counterfactual thinking in judgement and decision making and social psychology. Psychologists of reasoning have not yet even used what has been discovered in these other parts of psychology about how people think of some possibilities as closer to actuality than others. Much more integration in the field will deepen our psychological understanding of counterfactuals, conditionals more generally, causal reasoning, and decision making. And it could have significant relevance to normative debates about these topics. Our argument is that a necessary condition for this integration is a much deeper psychological understanding of the extended Ramsey test and of hypothetical thought.

# 8 Probabilistic accounts of conditionals

We believe that the Ramsey test, detailed in Chapter 2, provides an important insight into how people think about ordinary conditionals. As a reminder, the proposal is that people evaluate a conditional, 'if p then q', by making the antecedent p a hypothetical supposition and evaluating the consequent q in that context. This was a proposal made by a philosopher for good philosophical reasons. However, it is also a psychological proposal—with which we entirely agree—that 'if' triggers a process of hypothetical thinking. This proposal is consistent with a number of findings already discussed in this book, for example, the evidence that conditionals have a 'defective truth table' in the psychological literature (Chapter 3). Participants repeatedly tell us that 'not-p' instances have no relevance in determining the truth of a conditional.

The Ramsey test implies that people will evaluate the probability of a conditional, P(if p then q), as the conditional probability, P(q|p). We will refer to this implication as the *conditional probability hypothesis*. As a normative position, this hypothesis has been much discussed in the literature on philosophical logic, as we described in Chapter 2. A number of psychologists have also suggested, as an empirical hypothesis, that ordinary conditionals are related to conditional probability (George, 1995; Liu *et al.*, 1996; Oaksford and Chater, 1998, 2003b; Stevenson and Over, 1995, 2001). Curiously enough, however, only in very recent research have psychologists run experiments in which people are asked directly to assess the probability of conditional statements. Our primary purpose in this chapter is to introduce this experimental work, which is highly relevant, we believe, to the three broad theoretical accounts of the indicative conditional—T1, T2, and T3—that we introduced in Chapter 2 (Edgington, 2001, 2003). The strict truth of the conditional probability hypothesis would imply T3, but as we also explained in Chapter 2, and at the end of last chapter, it is not so easy to decide between T2 and T3. We will delay until chapter 9 our main discussion of the implications of the experimental work introduced in this chapter for T1, T2, and T3. However, we will begin with some basic theoretical points that we must be clear about.

## Clarification

Sometimes very loose statements are made about the relationship between 'if p then q' and P(q|p) that are no aid to clarity. For example, we have heard some psychologists speak as though the statement that P(q|p) = n is the same as statement that, if p holds, then the probability of q = n. Such a view is well known in philosophical logic as a fallacy, and has the absurd consequence that the only probabilities are 1 and 0 (Edgington, 1995, p. 269). It is also helpful to be clear about exactly what the conditional probability hypothesis says. It is the view that people evaluate P(if p then q) as P(q|p). It is not the claim, which we have heard some psychologists make, that 'if p then q' as a statement is 'equivalent' to P(q|p). Such a loose way of speaking makes no real sense, as the term 'P(q|p)' is not used to make a statement but is rather read as 'the conditional probability of q given p'. The Ramsey test and hypothetical thought more generally result in judgements of subjective probability, and thus in this context, P(q|p) is the conditional degree of belief in q given p.

Sometimes P(q|p), or the outcome of the Ramsey test, is held to equal the degree of belief that we would have in q if we *learned* p or came to *believe* p with certainty (Mellor, 1993). However, there are serious logical problems with this claim as well (Edgington, 1995, p. 270). There is a clear logical difference between assuming p and assuming that we believe or have learned p. There can also be quite a psychological difference in hypothetical thought between supposing that p and supposing that we believe or have learned p, whether with subjective certainty or not. We might hypothetically suppose that we were President of the United States. Under this supposition, we would infer that we could do something to slow down global warming. On the other hand, we might hypothetically suppose that we believed that we were President of the United States. Under this very different supposition, we would infer that we suffered from a serious delusion and should seek the help of one of our clinical colleagues.

Some researchers reformulate the original Ramsey test as part of a normative study of rational belief change (Gardenfors, 1998). The test then specifies that one should accept or believe 'if p then q' if and only if accepting or believing q would result from accepting or believing p. This reformulation could also be part of psychological study of how people's old beliefs do change as they acquire new beliefs. However, the Ramsey test was originally about what followed from supposing the antecedent of a conditional, and not about what followed from belief in the antecedent of the conditional. The original Ramsey test and the reformulated Ramsey test might often come to the same thing, but they can also arrive at very different places, as we have illustrated. Our stand in this book is founded on the original Ramsey test. We argue that it must be central to the explanation of recent experiments on the probability of conditionals, which we will now review.

## Judging the probability of conditional statements: the experimental evidence

Hadjichristidis *et al.* (2001) ran an experiment in which participants in separate groups were asked to make one of three different judgements about the same set of conditional statements, illustrated by the following examples:

*Probability of the conditional*
   Peter said the following: If horses have stenezoidal cells, then cows will have stenezoidal cells. How likely do you think it is that what Peter said is true?
   0   1   2   3   4   5   6   7   8   9   10
   not at all likely             very likely

*Conditional probability*
   Suppose you knew that horses had stenezoidal cells. How likely would you think it was that cows have stenzoidal cells?
   0   1   2   3   4   5   6   7   8   9   10
   not at all likely             very likely

The mean ratings for each condition were then correlated across the set of conditional sentences. Ratings between judgements of the probability of the conditional and those of conditional probability were very highly correlated (0.99). While providing evidence for the conditional probability hypothesis, there are inherent limitations in the correlational method. For example, if one set of judgements was consistently higher in magnitude than the other, a high correlation could still be obtained.

Two recent studies have provided very extensive experimental tests of the conditional probability hypothesis (Evans *et al.*, 2003a; Oberauer and Wilhelm, 2003). These studies were designed and conducted independently and yet used similar methods and drew similar conclusions. Evans *et al.* (2003a) designed their experiments to distinguish three empirical hypotheses about the basis on which people might judge the probability of an ordinary indicative conditional: the conditional probability hypothesis, the material conditional hypothesis, and the conjunctive probability hypothesis. The material conditional hypothesis is at the heart of the T1 family of accounts of the ordinary conditional, which we introduced in Chapter 2. It is an important hypothesis to investigate because psychologists have so often assumed, sometimes completely uncritically, that the ordinary conditional is the material, or truth functional, conditional.

Both Evans *et al.* (2003) and Oberauer and Wilhelm (2003) consider the predictions that the mental model theory of conditional reasoning (Johnson-Laird and Byrne, 2002) might make about these experiments, as this is the single most influential theory in the psychological literature. As we saw in Chapter 4, the theory proposes that 'basic' conditionals—of the type used in these experiments—have a 'core' meaning equivalent to the material conditional. However, the theory also states that people often have incomplete initial models of the conditional, in which the pq case is explicitly represented as well as three dots for the remaining, implicit mental models. It is unclear what can be predicted, from this proposal, about people with only initial models of the conditional. Will they assign the conjunctive probability, P(pq), as the probability of the conditional, or will they assign a rather higher probability to account for the three dots? Our view is that this version of mental model theory is committed to the prediction that people will assign a probability to the three dots. These dots are supposed to distinguish a conditional from a conjunction, and that is what they should do if they are present. Johnson-Laird and Byrne (2002) are not precise about a probability for the three dots, but they do clearly reject the hypothesis that a significant number of people will give the conditional probability as the probability of the conditional. Evans *et al.* and Oberauer and Wilhelm were also clear in their view that there is no psychological basis in standard mental model theory to support the conditional probability hypothesis.

The three probabilities computed by Evans *et al.* (2003a) are best illustrated with reference to one of their experimental problems:

A pack contains cards that are either yellow or red and have either a circle or a diamond printed on them. In total there are:

1 yellow circle

4 yellow diamonds

16 red circles

16 red diamonds

How likely are the following claims to be true of a card drawn at random from the pack?

If the card is yellow then it has a circle printed on it:

Very unlikely   1   2   3   4   5   Very likely

If the card has a diamond printed on it then it is red:
Very unlikely   1   2   3   4   5   Very likely

Ignore for the moment the second statement given, to which we will return. It is worth noting some points about the how the question was asked. First, the question was phrased to avoid any possible criticism of prompting for the conditional probability. According to Johnson-Laird and Byrne (2002), a problematic phrasing would have been, 'How likely is it that the card has a circle printed on it if it is yellow?' Second, the question was not pragmatic and about whether it would be appropriate or 'felicitous' to *assert* the conditional. (See Grice, 1989, and Jackson, 1987, for arguments about the assertability of conditionals, and Edgington, 1995, for criticism of this pragmatic view.) Third, the question referred to *the* single card drawn at random from the pack. Clearly the statement is false *of the whole pack* as this contains counterexamples. The frequency distribution provides the relevant information needed to compute the probability of the conditional. But what should this be?

- *Material conditional*: the material conditional is truth functional and is true unless the drawn card is a TF instance (antecedent true, consequent false), i.e. a yellow diamond. Hence, for the above example, the probability of the material conditional is 1–4/37, or 33/37.

- *Conditional probability*: under the conditional probability hypothesis, we need only consider the ratio of TT and TF cases. In the example, red cards are irrelevant. The conditional only 'applies' to yellow cards. Hence, the probability should be 1/5.

- *Conjunctive probability*: this sets the probability of the conditional as equal to the probability of the TT case. In this case the probability equals 1/37.

Note that in this example, the three probability calculations come up with very different answers. The second conditional statement included on each problem (Experiments 1 and 2 of the paper) was the contrapositive to the first. Under the material conditional hypothesis, the contrapositive must have the same probability as it shares the same truth table (only false in the case TF, otherwise true). However, under the conditional probability hypothesis, the probabilities of the conditional and its contrapositive can be very different. (See Chapter 2 on contraposition and the ordinary conditional.) In this case, the conditional probability for the contrapositive is 16/20, very much higher than the 1/5 chance for the conditional. In the experiments, Evans *et al.* used a number of problems with different frequency distributions so that all of these computed probabilities varied widely.

The first two experiments reported by Evans *et al.* (2003a) found evidence decisively against the material conditional hypothesis. For example, there was a near zero correlation between people's ratings of the probability of conditionals and their contrapositive statements. Evidence was more equivocal between the other two hypotheses. Both conditional probability and conjunctive probability correlated highly with the judged probability of the conditionals. In Experiment 3 of this paper, two different pack sizes were specified for two groups. It was so designed that the same set of conditional probabilities were involved in the materials given to both groups, but that the conjunctive probabilities were halved for groups with the large pack size, which had 60 rather than 30 cards.

For example, suppose the number of pq cards specified on a given problem was 6. Conditional probabilities could be varied by having problems with 2, 4, 6, 8, 10, or 12 p¬q cards (that is 6/8, 6/10, 6/12). The number of ¬p cards was filled out to the required pack size. Hence, for a small pack size, the conjunctive probabilities (of pq) would be 2/30, 4/30, 6/30, etc., whereas for the large pack size they would be 2/60, 4/60, 6/60, etc. By this method, Evans *et al.* were also able to show a strong effect of conditional probability with conjunctive probability controlled. The overall mean judgements are summarized in Figure 8.1. (This experiment used direct estimation of percentage probabilities.) It can be seen that in compliance with the conditional probability hypothesis, judgements increase (more or less linearly) with increased pq

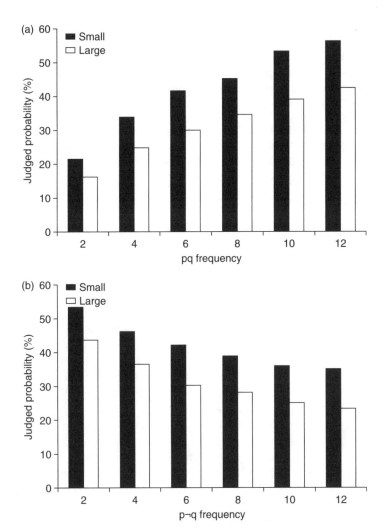

**Figure 8.1.** Judged probability (%) of conditionals as function of frequency of case and size of card pack (small/large) in the study of Evans *et al.* (2003).

frequency, and similarly decrease with descending p¬q frequency. However, there was still quite evidently an effect of conjunctive probability as evidenced by the substantial effects of pack size in the same graph. If people were judging only by conditional probability, pack size should make no difference.

Evans *et al*. (2003a) then investigated the possibility that their findings reflected individual differences. Suppose some people are using conditional probability but others are using conjunctive probability? Individual difference analysis did indeed reveal such differences. There was a clear subgroup of about 50% of participants who relied on conditional probability and completely ignored pack size. However, about 43% of participants instead relied on conjunctive probability and were highly sensitive to pack size. Findings of clear qualitative individual differences of this kind are most unusual in the thinking and reasoning literature.

Oberauer and Wilhelm (2003) replicated these findings in essential respects. Their experiments also used basic (abstract, indicative) conditionals, and they were equally careful about how they asked for probability of the conditional statement. However, their questions referred to samples rather than one-off probabilities. They also controlled their experimental tasks such that conditional probability and conjunctive probability were varied independently. Thus they had four groups of problems:

| | | |
|----|------------------------------------------------------------|------|
| HH | High conjunctive probability; high conditional probability | 82% |
| HL | High conjunctive probability; low conditional probability  | 51% |
| LH | Low conjunctive probability; high conditional probability  | 75% |
| LL | Low conjunctive probability; low conditional probability   | 41% |

The percentages shown above are the mean estimated probabilities given by participants in one of their experiments (1a). It can be seen that conditional probability substantially affects judgements, but conjunctive probability also affects them to a lesser degree. Oberauer and Wilhelm also found qualitative individual differences similar to those reported by Evans *et al*. (2003a). About 50% of their participants consistently matched the judged probability of the conditional to the conditional probability. A rather smaller subgroup than that observed by Evans *et al*. consistently used conjunctive probability. Their findings, like those of Evans *et al*., were also completely incompatible with probability judgements derived from the material conditional.

These two studies, taken together, provide very clear and consistent evidence. The dominant, but by no means exclusive, response is to assign the conditional probability as the probability of the conditional, as we might expect if people conform to the Ramsey test. The minority tendency to give the conjunctive probability might be claimed as some evidence for the mental models account. But standard mental model theory has to explain why the three dots do not get a probability, and of course it cannot account for the major trend, the conditional probability judgements. There are other explanations of the conjunctive response, which are consistent with the conditional probability hypothesis, as we will see below and in Chapter 9. Recall as well that the findings described to date use 'basic' or abstract indicative conditionals. In terms of the Ramsey test, these are highly impoverished in content, giving people little opportunity to supplement their hypothetical thinking by reference to relevant prior belief and knowledge.

Oberauer and Wilhelm (2003) extended their study to the case of thematic conditionals in their Experiments 3 and 4. However, these were arbitrary thematic conditionals that only make sense in the scenarios given. Their believability again depended upon explicit frequency information as in the experiments of Evans *et al.* (2003). A quite different approach was taken by Over *et al.* (in preparation). They used real world conditionals about which participants had genuine prior beliefs (see also Over and Evans, 2003). All the statements were causal indicative conditionals pertaining to events that might or might not occur in the near future. Participants were again asked about the probable truth of these conditionals. An example was:

If the cost of petrol increases then traffic congestion will improve

No frequency information was presented in this experiment. Instead, the relevant probabilities were computed from a probabilistic version of the truth table task. Participants (Experiment 1) were asked in one condition to rate the probability of each of the four cases for each of a set of 32 sentences, as follows:

How likely is it that:

| | |
|---|---|
| Petrol cost increases and traffic congestion improves | — |
| Petrol cost increases and traffic congestion does not improve | — |
| Petrol cost does not increase and traffic congestion improves | — |
| Petrol cost does not increase and traffic congestion does not improve | — |
| | 100% |

They were required to make their judgements of these TT, TF, FT and FF cases add up to 100%. From these ratings the relevant probabilities, such as P(q|p), can be computed. Note, however, that no mental arithmetic is necessary as no frequencies are presented. In a separate task, the participants rated the probability that the corresponding causal conditionals were true (presented in a different random order). From the truth table task, the following subjective probabilities were computed:

$$P(p) = TT + TF$$
$$P(q|p) = TT/(TT + TF)$$
$$P(q|\neg p) = FT/(FT + FF)$$

Across the 32 sentences, Over *et al.* then performed a multiple linear regression analysis using these three probabilities as predictors and the judged probability of the conditional as the dependent variable. This enabled Over *et al.* to tell whether one or more of these probabilities is affecting the perceived probability of the conditional and to what extent. These probabilities were chosen first of all because they are statistically independent measures. We can derive the following predictions:

*Conditional probability hypothesis*: P(q|p) should positively predict judgements
*Conjunctive probability hypothesis*: P(p) and P(q|p) should both positively predict judgements, as their product is equal to P(pq).

We also measured P(q|¬p) because these are causal conditionals, and a necessary condition for the existence of a causal relation is given by the $\Delta p$ statistic that we discussed in Chapter 7:

$$\Delta p = P(q|p)\text{---}P(q|\neg p)$$

The $\Delta p$ statistic, or 'delta p rule', measures the extent to which p *raises* the probability of q, when P(q|p) > P(q|¬p), and p must clearly raise the probability of q if p causes q. If participants are using the $\Delta p$ statistic to judge the truth of the conditional, then P(q|p) should predict the probability of the conditional positively, but P(q|¬p) should predict it negatively. We should get this result if a causal conditional is not only justified by a causal relation, but states semantically that the relation exists. The actual results for Experiment 1 were that conditional probability, P(q|p), predicted judgements of the probability of the conditional very strongly (beta weight = 0.90) while neither of the other probabilities had a significant effect. Contrary to the conjunctive probability hypothesis, the beta weight for P(p) was near zero. P(q|¬p) did predict negatively (−0.11), but the effect was small and non-significant.

It is possible that the $\Delta p$ statistic had no effect because people were asked to judge the truth of the conditional, which may not be the same thing as its perceived causal strength. This was checked in a replication experiment in which judgements of causal strength as well as the probability of the conditionals were made. P(q|p) predicted both sets of judgements very strongly (0.93). Again P(p) had no effect, but this time P(q|¬p) had significant effects on both judgements, although it was relatively small (−0.20 for probability judgements, −0.24 for causal strength judgements). The correlation between judgements of probability and causal strength was extremely high (0.98) in this experiment.

The findings of Over et al. are stronger than those of the studies of abstract conditionals discussed earlier. Since no frequency information is presented, the strong association with conditional probability cannot be explained as due to an artificial task inviting mental arithmetic. It is powerful evidence indeed that people are using the Ramsey test. What may well be an artefact in the studies of Evans et al. (2003a) and Oberauer and Wilhelm (2003) is the subgroup response based on conjunctive probability. There was no indication that this was a factor in the study of Over et al., as P(p) had no effect in either experiment. Working memory difficulties with mental arithmetic in the earlier experiments may have caused some participants to respond with the conjunctive probability (Over and Evans, 2003). The lack of an effect of P(p) also rules out the material conditional hypothesis, by which P(p) must predict negatively, as the material conditional increases in probability as the probability of the antecedent declines.

A further study using similar methodology has been conducted by Evans et al. (in preparation). They used similar 'causal' indicative conditionals referring to real world events but included negatives in the antecedent, consequent, or both. In a second experiment, they used counterfactual conditionals such as:

If Queen Elizabeth had died last year, then Prince Charles would have become King
If New York had not been attacked by terrorists in 2001, then the US would not have attacked Iraq

It is easy enough to ask people to judge the probability of the counterfactual conditional but trickier to obtain the probability of the truth table cases. Participants were asked to judge the probability that events might have happened relative to an earlier point in time, 5 years previously. Hence, they were to judge such probabilities as:

New York would be attacked by terrorists in 2001 and the US would attack Iraq
New York would be attacked by terrorists in 2001 and the US would not attack Iraq

Judging such probabilities in retrospect could be susceptible to hindsight bias (Fischhoff, 1982; Hawkins and Hastie, 1990; Baron, 2000; Roese, 2004), which might,

**Table 8.1.** Regression weights (across sentences) for three predictors in the study of Evans *et al.* (in preparation)

|  | Probability | Causal strength |
| --- | --- | --- |
| Indicatives |  |  |
| P(p) | 0.04 | 0.07 |
| P(q|p) | 0.95* | 0.91* |
| P(q|¬p) | −0.28 | −0.35* |
| Counterfactuals |  |  |
| P(p) | 0.06 | 0.13 |
| P(q|p) | 0.87* | 0.85* |
| P(q|¬p) | −0.34* | −0.34* |

for example, lead people to overestimate the likelihood of an event that they know happened. However, this bias should not confound the hypothesis of interest to us, namely whether the conditional probability derived from such beliefs will correspond with the perceived probability of the whole conditional. Negations had little effect on judgements, so we summarize the results of this study in Table 8.1 for all sentences, again reporting multiple regression analyses on the mean ratings of participants across sentences. What is very striking about the findings is that results are very similar for indicative and counterfactual conditionals and for ratings of probability or of causal strength. To our knowledge this is the only psychological experiment that has so far examined the perceived probability of counterfactual conditionals. The strong predictor in all cases, in line with the Ramsey test is the conditional probability P(q|p). However, there is also a secondary negative effect of P(q|¬p) in line with the delta p rule.

The limited effects of P(q|¬p) in this study and that of Over *et al.* described previously is evidence for the strength of the Ramsey test as a psychological hypothesis. Under the delta p rule it should have an effect equal in strength to P(q|p). Clearly, the false antecedent case is relevant to the strength of a causal relation. For example, it is not enough to establish a causal relation to find out that traffic congestion eased when the price of petrol went up. One needs to show that the petrol price 'intervened' in some way to change the world (Pearl, 2000). If traffic congestion would have eased without petrol prices going up, then it could be quite unrelated, and responsive to some other cause, e.g. more road building. In spite of high causal relevance, people's actual judgements—of causal strength as well as truth—are clearly much more affected by P(q|p) than by P(q|¬p), as measured from their own prior beliefs about the relevant events. This result can be explained by the Ramsey test, which focuses attention only on possibilities in which the antecedent holds. It is also consistent with our theory of hypothetical thinking (Evans *et al.*, 2003b) in which we propose that people generally consider only one possibility at a time.

The findings of these experiments are also consistent with much work on people's judgement about causal relations and the 2 × 2 tables we described in Chapter 7. This work shows that people's causal judgements tend to be more strongly influenced by entries in the a and b cells than in the c and d cells. The former correspond to the frequencies needed to compute P(q|p) and the latter those needed for P(q|¬p). There has been much discussion about why this tendency exists and the extent to which it is a bias or rational in many contexts (Anderson, 1990; Shanks, 1995; Stanovich and West, 1998b, 2003; Over and Green, 2001). It might be that a 'causal' conditional does not semantically state the existence of a causal relation. In that case, using a high P(q|p) alone, or just information from the a and b cells, to evaluate a causal conditional would not be a bias. That would make the term 'causal conditional' possibly misleading for these conditionals. On the other hand, causal conditionals might semantically state that a causal relation exists. Then using P(q|p) alone, or just information from the a and b cells, could be a bias, perhaps caused by our tendency to focus on only one possibility when engaged in hypothetical thinking. This issue is bound up with the problem of whether there is a special class of indicative conditionals that are closely tied to counterfactuals (see Chapters 7 and 9).

## Probabilistic treatments of conditional inference

The research reviewed above provides solid evidence for the role of the Ramsey test in the probabilistic evaluation of conditional statements. This is not quite the same thing as the probabilistic *representation* of conditionals that some psychologists have proposed. Certainly a conditional 'if p then q' must be represented as something more than simply P(q|p). As we pointed out at the beginning of this chapter, 'P(q|p)' does not stand for a proposition, but rather refers to a conditional degree of belief in q given p. Conditional statements vary in their degree of *assertability* (Edgington, 1995), whereas conditional degrees of belief do not. And people do not consider the conditional probability to be the only basis for asserting a conditional, even if they judge the probability of the conditional to be the conditional probability. Nor do we believe that the hypothetical thought process that occurs in the Ramsey test is confined to judgements of probability.

Consider the following two statements of conditional advice from one student to another:

8.1  If you revise these three topics then you will pass the exam
8.2  If you revise all the topics then you will pass the exam

Statement 8.1 seems assertable provided that the probability of q given p is high. Statement 8.2 appears less assertable, even though the conditional probability must be at least as high, and probably a lot higher. Most students would reject this advice as unhelpful. Why? The answer is because the cost of the antecedent action is too high. Statement 8.1 would also not be relevant advice to a student who did not need to pass the exam—say they had just been offered a lucrative job that did not require completion of the course in question. Thus the mental simulation engaged by the conditional involves consideration of costs and benefits as well as of probability. Conditional advice may even be acceptable when P(q|p) is quite low, where someone is desperate to achieve q knows of no other way of doing so, as in someone dying of a disease who tries an unproven treatment. Oberauer and Wilhelm (2003, Experiment 3) actually showed that people's estimation of the probability of the conditional was influenced by the desirability of the outcome specified by q.

So granted that there is a good deal more to conditionals and hypothetical thought than judgements of conditional probability, let us examine some of the psychological work that treats conditionals probabilistically. In Chapter 3, we noted some relevant work by Evans *et al.* (1996b) in which people were asked to construct distributions of truth table cases to represent true and false conditionals (see Figure 3.1). The results were suggestive of a probabilistic representation, in that participants included a small number of counter-example TF cases when modelling true conditionals and a large number when modelling false conditionals. This suggests that they took the 'true' and 'false' instructions to represent extreme cases when P(q|p) was very high and very low, respectively. Of course, on a logical interpretation they should include no TF cases at all for true conditionals, and need only have included one (not many) when modelling false conditionals. There could, however, be an 'interpretation' problem here. Participants might have treated a general conditional, e.g. 'if the symbols are circles then they are yellow', as specific, e.g. 'if a specific circle is picked out then it will be yellow'.

Most of the relevant work has used the conditional inference paradigm already discussed in detail in this book, especially in Chapters 3 and 6. (Probabilistic treatments of the Wason selection task are discussed later in this chapter.) In Chapter 6, we reviewed several studies on thematic conditional inference that suggested probabilistic processing (George, 1995; Stevenson and Over, 1995; Dieussaert *et al.*, 2002). In these studies, participants were allowed to express degrees of belief or confidence in pro- posed conclusions to conditional arguments and happily did so, rather than sticking to the extremes of the scale, as logical processing would require. Moreover, these studies showed that the degree of belief in the conclusions was related to belief in the condi- tional premise. We also reviewed experiments that showed that the rated degree of per- ceived necessity and sufficiency of p for q with particular thematic conditionals affected willingness to draw inferences. The study of Liu *et al.* (1996) obtained sufficiency and necessity ratings by directly asking participants to rate the probabilities P(q|p) and P(p|q). These values predicted both (1) proportion of conclusions accepted when a determinate (forced-choice) response task was used, and (2) mean confidence in con- clusions when a probabilistic response task was used.

Oaksford *et al.* (2000) put forward an impressive probability model of conditional inference in which they applied probability theory. This model has three parameters, a = P(p), b = P(q), and $\varepsilon$, the exceptions parameter. The latter is the rate of exceptions that people will tolerate when believing the conditional statement, that is $1 - P(q|p)$. Using these parameters, Oaksford *et al.* compute the probability of the four conditional inferences by straightforward application of the probability calculus as follows:

MP   $1 - \varepsilon$
DA   $(1 - b - a.\varepsilon)/(1 - a)$
AC   $a(1 - \varepsilon)/b$
MT   $(1 - b - a.\varepsilon)/(1 - b)$

In conditional arguments tasks, participants are given a major premise (the condi- tional) and minor premise and a putative conclusion to evaluate. The Oaksford *et al.* (2000) model assumes that the perceived strength of the argument is directly related to the conditional probability of the conclusion given the minor premise. For example, confidence in the conclusion of MT for the conditional 'if p then q', should be equal to P(not-p|not-q). It also assumes that the normatively calculated value, as shown above, will predict performance. There is no apparent psychological mechanism proposed to account for this, but the theory is consistent with Oaksford and Chater's rational analysis research programme (Oaksford and Chater, 1998), which assumes that behaviour approximates a normatively justified standard. Effectively, they have substituted the traditional normative model of propositional logic with that of probability theory.

Oaksford *et al.* (2000) offer experimental psychological evidence for their model, by asking people to draw inferences from conditional statements that are perceived to be high or low in P(p) and high or low in P(q). One supportive finding was that people tended to endorse more inferences whose conclusions were computed to have high probabilities under the model. This could be related to the belief-bias effect discussed in Chapter 6. We know from that literature that people tend to favour conclusions that have prior believability, in spite of instructions to reason deductively. However, Oaksford *et al.* failed to find evidence for predicted effects of premise probability.

Close inspection of Oaksford *et al.*'s paper reveals something rather odd. In illustrating the behaviour of their model (their Figure 1) they set the exceptions parameter, $\varepsilon$ to 0.75. This value would seem to be much too high, as MP rates observed in psychological experiments (abstract materials, adult participants, standard reasoning instructions) typically approach 100% unless special manipulations are introduced, as revealed in earlier discussions in this book. The difficulty for their model is that observed rates of MT are typically moderate (60–75%), whereas a low value of $\varepsilon$ forces them to be high. In the extreme case, where $\varepsilon = 0$, MT rates, like MP rates would be 100% as they are both valid inferences. We estimated $\varepsilon$ from the study of Evans *et al.* (1996) by looking at the number of exceptions (TF cases) people included when modelling 'true' conditionals. Two different experiments yield extremely similar estimates of $\varepsilon = 0.067$ (Experiment 2) and 0.066 (Experiment 3) respectively, which would predict MP rates of about 94%, actually an underestimate of what is typically observed. Even so, this presents real difficulties for the model in predicting MT rates for abstract conditionals; say the 73% observed by Evans *et al.* (1995). In our own investigations we found it impossible to fit this rate with b set to less than 0.8. In other words, with low exceptions, q must be very probable in order to account for the MT rates that we see in abstract reasoning experiments. However, in fitting their model of the abstract selection task, Oaksford and Chater (1994) argue in contrast (and quite convincingly) that in such experiments people assume by default that both p and q are *rare*. The selection task model can only be fitted by setting P(q) to a low, rather than high value.

The rarity assumption was discussed in an exchange between Schroyens and Schaeken (2003) and Oaksford and Chater (2003a) as the former authors noted the same apparent contradiction between the two models as we note above. In reply, Oaksford and Chater argue that the a and b probabilities are overestimated in the conditional inference task because these inferences are only relevant if the events are more likely than normal to occur. In other words, the pragmatics of the task overrides the default rarity of p and q. We find ourselves unconvinced by this argument. There is indeed a pragmatic factor at work that implies fairly high probabilities for p and q, but it lies in the assertion of the conditional statement itself, rather than the requirement to draw conditional inferences from it. 'If p then q' is not assertable—or at least has low relevance—in most contexts if P(p) is too low or if P(q|p) is too low. As conditionals only apply to p-states, such states must normally be reasonably probable (at least in the near future) for a conditional statement to have relevance. Similarly, the conditional is only assertable when P(q|p) is high, because this is the link assumed, which implies for high P(p) that P(q) is also high. Unfortunately, for Oaksford and Chater, however, precisely the same argument would apply to the assertion of a conditional on the Wason selection task.

While the evidence of conditional reasoning studies clearly supports the view that people process conditionals in a probabilistic manner, we should make it clear that our own theory does not follow the direction set out by Oaksford and Chater. First, as stated earlier, we believe that there is more to the mental representation of a conditional statement than conditional probability. Probabilistic evaluation is simply one consequence of hypothetical thought about conditionals. Second, a psychological account of the mechanisms of conditional inference must be given that goes beyond computing conditional probabilities. We shall discuss such an account in Chapter 9.

## Probabilistic accounts of the Wason selection task

The Wason selection task, discussed in detail in Chapter 5 seems to be a curious mixture of a reasoning and decision-making task. The task instructions require people to make a choice, to select cards. The conventional normative analysis, presented by Peter Wason in his original studies, assumes that conditional logic provides the basis for this choice. However, we have already seen in Chapter 5 that choices may be strongly influenced by the perceived relevance of the cards and that the processes underlying such perceptions of relevance have more to do with expected benefits and costs than any process of logical reasoning. The decision theoretic approach, which we explored in some detail in an earlier book (Evans and Over, 1996) attempts to understand behaviour on the task by assuming that people's choices are influenced by personal goals that may be either epistemic (abstract task) or practical (deontic task). In this case, probabilities and utilities come into play.

To our knowledge, the earliest study to examine the effects of degree of belief in the conditional statement, in the context of the selection task, is that of Van Duyne (1976). Van Duyne suggested that people were motivated by 'cognitive self-reinforcement' such that they would attempt to confirm a rule that they believed to be true, but to refute a rule that they believed to be false. There are connections here with recent theories of belief bias, discussed in Chapter 6 (Klauer *et al.*, 2000; Evans *et al.*, 2001). Van Duyne's experiments were criticized by Pollard and Evans (1981) who nevertheless found evidence in their own experiments that was consistent with his hypothesis. In the Pollard and Evans study participants were asked to construct conditional statements that were: (1) always true; (2) sometimes true; (3) sometimes false; and (4) always false. They were then given selection tasks to solve with their own sentences. Choices frequencies were quite sharply different for true-consequent and false-consequent cards. Combining 'sometimes' and 'always' versions, they found higher choices of TC for true (92%) than false (79%) sentences, but lower choices of FC for true (61%) than false (87%) sentences.

The Pollard and Evans (1981) findings thus show apparently more disconfirmatory choices when the rule is believed to be false. In a later paper, Pollard and Evans (1983) produced a similar finding with abstract conditional statements, when the selection tasks were preceded by a probability learning task. Packs of cards were used in which p & not-q combinations were either rare or frequent and knowledge of these frequencies was established by experimental training. Four cards were then drawn from the pack in view of the participants with the visible sides of p, not-p, q and not-q. Thus, although participants had uncertainty as to whether the rule was true or false on each selection task, they knew that it was *probably* true or false based on their experience. As predicted, there were significantly more FC choices for false rules and more TC choices for true rules. The explanation favoured by Pollard and Evans was that the manipulation made a counter-example case of p & not-q available to the participants. The presence of such cases was explicitly demonstrated by the training method of Pollard and Evans (1983).

Explicit presentation of relevant frequencies was first introduced into selection task experiments by Kirby (1994) and produced results compatible with Pollard and Evans's account, in spite of some debate about the methodology used (Over and Evans, 1994). In his Experiment 2, for example, participants were told that a computer was given the task of printing out cards with an integer from 1 to 100 on one side, with one of two symbols (+ or −) on the other side. There were three conditions as follows:

- small p set—if a card has a 1 on one side then it has a+ on the other side
- medium p set—if a card has a number from 1 to 50 on one side then it has a+ on the other side
- large p set—if a card has a number from 1 to 90 on one side then it has a+ on the other side

Kirby predicted and observed that not-q selections were more frequent as p set size increased. The point is that as p-set size increases, so the probability of finding p & not-q cards in packs printed by this computer also increases. So, in turn, the likelihood of discovering a falsifying case by turning the not-q card in a set of cards increases. The difference between this study and that of Pollard and Evans (1983) is that it used explicit frequency information, rather than experience-based training. However, the findings are compatible. The condition that facilitated not-q choices in the Pollard and Evans experiments was one where the packs used for training and high numbers of p & not-q cases. These experiments show that participants are clearly sensitive to information about probability distributions on a task that purportedly requires only a grasp of the logic of conditional statements for its solution.

These findings are also compatible with accounts of the standard selection task that relate it to Hempel's paradox from the philosophy of science (Nickerson, 1996). This 'paradox' can be illustrated with the conditional claim:

8.3   If it is a raven then it is black

Clearly, one can confirm 8.3 by looking for ravens and checking whether they are black. However, in extensional logic, the above statement is logically equivalent to its contrapositive:

8.4   If it is not black then it is not a raven

Hence, it seems that one should equally look for non-black things to check whether they are non-ravens. Yet, this is obviously a hopeless strategy because the negative classes so defined are vast in size. Finding that a non-black object, such as a piece of chalk or a mute swan, is a non-raven is not going to increase measurable confidence in 8.3. One resolution of the 'paradox' is to point out that, in Bayesian inference, observing a non-black non-raven provides minimal support for 8.3 compared with observing a black raven. We should also remember that the ordinary conditional 8.3, according to T2 and T3, does *not* logically imply its contrapositive statement 8.4. Proof of contraposition rests upon the standard, extensional rule of if-introduction, which does not apply to the ordinary indicative conditional if T2 or T3 hold (see Chapter 2).

Oaksford and Chater (1994) developed a very stimulating probabilistic account of the abstract selection task using their rational analysis approach. They argued that standard responses on the task, normally classified as errors, could be seen as correct responses when an alternative normative framework based on expected information gain was applied. They called this the Optimal Data Selection or ODS model. The idea was that people's choices were determined by the amount of information they could expect to gain by choosing particular cards. The calculations presupposed a rarity principle: that people would assume by default that the p and q set sizes are small and so P(p) and P(q) would have relatively low values. One could question the rarity assumption. (In Hempel's paradox, the set of ravens is relatively small, but is the set of all black things?) But given rarity as a default assumption, people would expect to learn little by turning over the not-q card. The details of Oaksford and Chater's model and their assessment of evidence in its favour proved somewhat controversial (Evans and Over, 1996b; Laming, 1996; Oaksford and Chater, 1996; Klauer, 1999). Subsequent research with probability manipulations has resulted in conflicting claims about the extent to which the ODS model is supported (Green *et al.*, 1997; Oaksford *et al.*, 1999; Oberauer *et al.*, 1999).

It is not necessary for our current purposes to examine the details of the ODS model or the controversy surrounding it. What is relevant is to note that there are number of experiments now published that show participants are sensitive to probabilistic manipulations on the indicative selection task. Broadly speaking, these show that people are more likely to choose the not-q card when they have reason to think that the rule may be false and there is a reasonable chance of finding p on the other side. This result is consistent with the findings of many other experiments on both indicative and deontic selection tasks, which imply that correct selections result from attention being directed to the counter-example or violating case of 'p & not-q' (see Chapter 5).

## Conclusions

In this chapter, we reviewed further support for the conclusion that the ordinary conditional of natural language is not the material, truth functional conditional. We also discussed the growing evidence that people judge the probability of the ordinary conditional to be the conditional probability. In our eyes, the logical and philosophical arguments are very strong against the T1 claim that the ordinary conditional is the material conditional (see Chapter 2). We have shown in addition that the psychological evidence against this claim is overwhelming. The evidence of the direct studies of probabilistic assessment of conditionals presented by Evans *et al.* (2003a) and by Oberauer and Wilhelm (2003), in combination with studies using other methods reviewed earlier in this book, makes the material conditional hypothesis and T1 untenable. Contrary to the claims of the mental model theorists, Johnson-Laird and Byrne (2002), this applies equally to 'basic' (abstract, affirmative) conditionals, as to those embedded in pragmatically rich contexts. There are other psychological accounts (Braine and O'Brien, 1991), and some philosophical analyses (Lycan, 2001), of the ordinary indicative conditional that do not explain at all its relation to conditional probability (Over and Evans, 2003). As long as that is so, they are untenable as well, given the evidence that we have reviewed.

# 9    Towards a suppositional theory of *if*

This book is concerned with the ordinary conditional of everyday discourse. In our quest to understand it, we have examined both the contribution of philosophical logicians and psychologists to this important topic. We have reviewed the arguments of philosophers that the ordinary conditional 'if p then q' cannot be the material, truth functional conditional, which asserts no more than 'not-p or q'. This philosophical literature is the source of a sketchy hypothesis but one that we believe to have great psychological potential: the Ramsey test (Chapter 2). According to this hypothesis people evaluate the probability of a conditional by hypothetically supposing that the antecedent p holds and assessing the believability of q under that supposition. This suppositional assessment or evaluation of conditionals avoids the major difficulties of the material conditional, including the well known paradoxes. We also introduced, in Chapter 2, the three families of theories—T1, T2, and T3—that philosophers have developed in their analyses of ordinary conditionals (Edgington, 2001, 2003). We will say more in this chapter about how the psychological evidence bears on T1, T2, and T3.

The psychological literature on conditionals is massive and the great majority of it focused on conditional reasoning within the deduction paradigm (Evans, 2002a). This paradigm, born in a period of strong 'logicist' thinking in psychology, is far from ideal for the study of the ordinary conditional, as we have argued in this book. The problem is that the standard instructions for these experiments tell people to alter the default mode of reasoning with conditionals. They are told to assume that the premises are true and to draw only conclusions that necessarily follow from these restricted assumptions. In the first place, these experiments do not even try to distinguish between what necessarily follows from *beliefs* or anyway premises that are more or less *probable*, and what necessarily follows from *assumptions*, which are to be *supposed true*. In all these standard experiments, people are asked whether arguments are valid and not whether the arguments are sound, where a sound argument is valid and has true premises. They are always to resist pragmatic inferences and to ignore uncertainty in the premises, a tendency that would be irrational in realistic reasoning. In contrast, completely realistic reasoning instructions should ideally allow people to introduce prior beliefs, make pragmatic inferences, draw conclusions that plausibly, probably, or necessarily follow, and express a degree of confidence in the conclusions.

Fortunately, there is a useful minority of studies in which instructions approach the ideal for studying ordinary conditional inference (see, for example, Chapter 6). We can also learn a great deal about ordinary conditionals from the majority of experiments within the deduction paradigm, but the interpretation is more complex. Pragmatics and assessment of uncertainty tend to assert themselves in spite of the instructions given, but understanding how such tendencies compete with a conscious effort at deduction requires the use of the dual process theory of reasoning to which we have alluded several times. We will return to this issue later in the chapter.

In summary, the psychological evidence strongly disconfirms the T1 claim that the ordinary indicative conditional is truth functional. It provides positive support for the Ramsey test (see, for example, the defective truth table, discussed in Chapter 2, and the topic of probabilistic evaluation, discussed in Chapter 8), and supplies a great deal of information about the effects of content and context on the representation of conditionals. People often appear to treat questions about validity as if these were about soundness, or about their confidence in conclusions that follow from uncertain premises. They automatically make pragmatic inferences to conclusions that may not logically follow, but that may be probable given the premises. They tend to be reluctant to draw conclusions from premises they disbelieve. Their confidence in an indicative conditional premise appears to be based on the conditional probability. More impressively still, people can report their degrees of belief in indicative conditionals, and we can measure their implicit conditional probability judgements. These judgements do tend to be at least close to each other. The evidence clearly supports the conditional probability hypothesis we discussed in Chapter 8. What all this evidence points to is a mental representation of a conditional, 'if p then q', that includes a subjective connection between the representation of p and that of q, indicating the degree of belief in q given p.

The dominant theory and paradigm in the psychology of reasoning is undoubtedly that of mental models, led by Johnson-Laird and his collaborators (Johnson-Laird, 1983; Johnson-Laird and Byrne, 1991). They originally took a T1 view of indicative conditionals in natural language. As we have indicated in Chapter 4 and elsewhere in this book, we believe that the most recent account of conditionals within this paradigm (Johnson-Laird and Byrne, 2002) is mistaken in several significant and important ways (see also Evans *et al.*, 2003c; Handley and Feeney, 2003; Over and Evans, 2003). Understanding why this is so is the key to the development of better psychological account of 'if'. For this reason, and because of the great popularity of this approach in the psychological literature, we start with some brief further discussion of this approach.

## Mental models

The term 'mental models' is ambiguous in several ways. First, the term can be used in a very general way that has little connection with the influential theory of Johnson-Laird and his collaborators. The term can simply refer to any kind of mental representation that describes or simulates a hypothetical or actual state of affairs. For example, use of the term to describe mental simulations (Gentner and Stevens, 1983) implies a dynamic mental model that might include rules or schemas embedded within it. Nothing of this sort has been proposed in mental model accounts of deductive reasoning, either by Johnson-Laird or anyone else.

A second, more specific use of the term 'mental model' lies within the general theory of mental models that Johnson-Laird and sympathetic researchers have developed. Here, the models are static, simply representing logical possibilities. The general theory makes proposals that many psychologists have found attractive. Deductive competence is determined by a semantic principle, not relying on inference rules or natural logics, as proposed by an older tradition which is still supported by some authors (Rips, 1994; Braine and O'Brien, 1998a). The psychological account of deductive effort given is that people build mental models of the premises, from which they form putative conclusions. They then attempt to validate these conclusions by searching for counter-examples: mental models in which the premises hold but not the conclusion.

This theory is clearly psychological in nature, allowing competence in principle but proposing that it is difficult to achieve in practice. The problem is that people have limited working memory capacity and may find it difficult to represent more than one mental model, or to fully model non-trivial states of affairs. This theory was first applied to both syllogistic and relational reasoning with some success (Johnson-Laird and Byrne, 1991). However, it has run into some difficulties more recently. In particular, there is mounting evidence that people do not ordinarily engage in a search for counter-examples in syllogistic reasoning (Evans *et al.*, 1999a; Newstead *et al.*, 1999). It seems instead that people tend to build a single mental model of the premises and draw a fallacious inference if and only if their preferred model includes a conclusion that is possible but not necessary. Recent accounts of the belief bias effect in syllogistic reasoning have accordingly proposed that conclusion believability affects whether the single model people build is to confirm or refute the presented conclusion (Evans *et al.*, 2001; Klauer *et al.*, 2000). This revision to the general theory is consistent with our own principles of hypothetical thinking (Evans *et al.*, 2003b, see Chapter 1), which include the proposal that people tend to consider only one hypothesis at a time.

We would like to make it clear that we do not have a problem with mental model approaches in the first very broad sense, or even in the second sense of the general mental model approach to reasoning. Indeed, in general terms our own approach could be classed as a mental model theory. What we do have very serious problems with, however, is the specific account of conditionals given by the leading advocates of this approach, most recently updated by Johnson-Laird and Bryne (2002; see Chapter 4 for a detailed critique). It is worth reminding the reader of the basic problems that we have identified.

The Johnson-Laird and Byrne (JLB) 'basic' conditional is the material, truth functional conditional, and their T1 account of it is rooted in extensional logic. Johnson-Laird and Byrne (2002) endorse the validity of the paradoxes of treating the ordinary conditional as truth functional and argue against accounts of it in terms of subjective conditional probability. We have provided strong evidence in this book against this position. Not even the ordinary 'basic' conditional is truth functional. People's mental models of conditionals must include directionality and degrees of strength that are exhibited by measures of confidence, belief, and probability in both the conditionals themselves and the conclusions that people are willing to draw from them.

The JLB theory of conditionals does not appear to us to conform to all the principles of the general mental model theory of reasoning that these authors have so successfully developed. The general theory talks of validating inferences by searching for counter-examples. As we mentioned earlier in the book (for example, Chapter 6), some authors working in the mental model traditional have applied this concept to conditional inference, especially when accounting for pragmatic influences. However, our reading of the Johnson-Laird and Byrne (2002) paper, and their earlier papers, does not indicate any clear role for this mechanism in their specific theory of conditionals. First, they emphasize a 'principle of truth' in which only true possibilities are represented. There is hence no direct representation of false or counter-example cases. Initially, incomplete representations may be 'fleshed out' to a fully explicit list and it is here, we assume, that the mechanism of pragmatic modulation operates, for example, to omit logically possible cases that are precluded by the context. The inferences people draw are then a function of the model set (partial or fleshed-out) that is available. There is, so far as we can see, no specification of any further search for counter-examples to a putative conclusion from a conditional inference in the JLB theory, which would in any case provide rather a surplus of explanatory mechanisms. The literature seems confused on this point, however, as authors using the counter-example concept in their papers often cite Johnson-Laird and Byrne. Moreover, Byrne's own account of the suppression of Modus Ponens (MP) by additional premises (Byrne *et al.*, 1999), discussed in Chapter 6, *does* try to specify a representation of a counter-example case, rendering MP a supposedly invalid inference. We feel this is an area that needs sorting out and deprecate the common practice of psychologist to refer to *the* mental model theory or reasoning, as if there were only one.

Clarification would also be welcome on the semantics of conditionals. As we showed in Chapter 4, Johnson-Laird and Byrne (2002) have a semantics for 'basic' conditionals that must be truth functional, given their endorsement of the validity of the paradoxes of the material conditional. This semantics puts them in the T1 camp on basic conditionals. On the other hand, they refer to a mechanism of 'modulation' that is supposed to imply that non-basic conditionals are non-truth functional, because of the way real world knowledge is relevant to their truth conditions. They are far from stating a clear theory of these non-basic conditionals, which must be grounded in intensional semantics (Chapter 2) and not the present extensional semantics of JLB. Will the final mental model theory of these conditionals be in family T2 or family T3? To answer this question, we must know whether a non-basic conditional has a truth value when its antecedent is false. If so, then mental model theory should in T2. If it does not have a truth value when its antecedent is false, then mental model theory should be in T3. We admit that the question is hard to answer, as much for us as for them. But it would be progress if they acknowledged the necessity of an answer. From that point, they could try to develop the semantics of their non-basic conditional in more detail, and say something about its probability. Actually, as we pointed out in Chapter 4, they have an even more elementary problem, in that they deny that non-basic conjunctions and disjunctions are truth functional. They say nothing about the intensional semantics, as it would have to be, of these non-truth functional conjunctions and disjunctions, which obviously could not conform to extensional propositional logic nor to standard probability theory.

Modulation in JLB theory is unable to cope with much of the data we have discussed in this book. Although Johnson-Laird (personal communication, see Chapter 4) has suggested that it is broader, we believe that their principle of pragmatic modulation has only been well defined and applied within their extensional framework. That is, it appears to operate by adding or subtracting possibilities in the list provided by the 'core' semantics of the conditional: what we call the four-bit semantic device (Evans *et al.*, 2003c, see Chapter 4). This is a purely extensional semantics in which the meaning of all conditionals in all contexts can be reduced to just 16 possibilities. The extensional approach also means, as we said above, that p and q are not connected in any way in the mental models, and no degree of strength or probability between p and q can be represented. (See Over, 2004a,b, for more on mental model theory of probability and of conditionals.)

Our own mental model approach to conditionals is radically different. For us, the original sin of JLB theory is their endorsement of the logical validity of the paradoxes of interpreting a natural language conditional as truth functional. We hold that the paradoxes are logically invalid for all natural language conditionals, including 'basic' conditionals. People use the Ramsey test to assess conditionals, focusing on the antecedent possibility. As a result of applying this test in hypothetical thought, people can often construct mental models that are rich and dynamic, representing far more than just a logical possibility or a truth table case. We say more about this later in the chapter. As indicated in Chapter 1, our theory has three foundations: the Ramsey test, pragmatic inference, and dual process theory. We now elaborate these in detail and provide an account of the major phenomena reviewed in this book in the process.

## Supposition and the Ramsey test

The Ramsey test, strictly speaking, is too narrow for our purposes. The test, as originally stated by Ramsey, is a way of arriving at a degree of belief in a conditional. We are to do this by judging the extent to which we believe q given p. The evidence is strong that this is how people assess or evaluate the believability of the conditional, but the hypothetical or suppositional approach that underlies it has much broader application. It is the suppositional nature of conditional thought that is the primary concept: the Ramsey test should be viewed as only one of its important psychological consequences.

As we said in Chapter 1, our view of 'if' is that of a linguistic device the purpose of which is to trigger a process of hypothetical or suppositional thinking and reasoning. We believe hypothetical thought to be a unique feature of human cognition: no other animal, we believe, entertains a hypothesis, or makes a decision by imagining and evaluating consequences in possibilities. This is why the understanding of 'if' is not a narrow academic concern, but a matter of central importance in the understanding of what makes human intelligence special and distinctive.

Let us recap some of the experimental evidence that has been presented in this book for the suppositional nature of 'if'. The most direct evidence comes from two main sources: studies of how people judge the truth value of conditionals in truth table tasks (Chapter 3) and studies that ask people to assess the probability of conditional statements (Chapter 8). The major trend in truth table tasks is for people to indicate a 'defective' truth table in which not-p instances are described as irrelevant to the truth of the conditional. We interpret this finding as evidence against T1 and supportive of the Ramsey test. The truth of a conditional is always decided by reference to p instances, and not-p instances are classified as irrelevant. However, we must be cautious here in claiming that these results are evidence for T3, which states that an indicative conditional does not have a truth value when its antecedent is false. The conditionals used in truth table tasks were quite abstract general conditionals, for example, about cards in a pack with letters and numbers on them. A supporter of T2 might be no more impressed by these tasks than by an experiment in which participants classified a mute swan as irrelevant to the claim that all ravens are black. (See Chapter 8 on Hempel's paradox.)

As we pointed out in Chapter 3, even if T3 were the best account for abstract conditionals in truth table tasks, the supporter of T2 could argue that T2 was much better suited to highly thematic conditionals. There is no intuitive notion of closeness between possibilities in truth table tasks performed with abstract conditionals, but there can be for more realistic conditionals.

Potentially more far reaching evidence comes from experiments on the judged probability of conditionals. As we saw in Chapter 8, the majority of people judge the probability of 'if p then q' to be at least close to the conditional probability P(q|p). This is again precisely what we would expect if people were conforming to the Ramsey test. We also saw, in Chapter 6, the consequences of uncertainty in the premises of conditional inferences, especially the valid inferences Modus Ponens (MP) and Modus Tollens (MT). There is evidence that confidence in the conclusions of these inferences, when the major premise is uncertain and the minor premise certain, is determined by the conditional probability P(q|p). As we explained in Chapter 2, if the probability of an ordinary conditional is precisely the conditional probability, then T3 is confirmed. However, as we also argued in Chapter 2, there are reasons for thinking that, even if T2 holds, the probability of an ordinary conditional may often be close to the conditional probability (see also Chapter 7).

There are significant minority trends in these experiments for which we need to account as well, and these take us beyond the Ramsey test and into pragmatics. In truth table tasks, quite a lot of people declare that the 'not-p & q', FT instance is not irrelevant but makes the conditional false. This response probably results from a biconditional interpretation and pragmatic inference, and we discuss this possibility in the following section. When people are asked to judge the probability of relatively abstract conditionals about frequency distributions (Hadjichristidis *et al.*, 2001; Evans *et al.*, 2003a; Oberauer and Wilhelm, 2003; Over and Evans, 2003) a substantial minority give the probability of the conjunction P(p & q). It clearly cannot be the case that intelligent adults really think of conditionals as conjunctions, as they would be intellectually disabled and incapable of hypothetical thinking. Regression analysis shows that the conjunctive response is not a factor in assessing more realistic conditionals, and so some people may find rather abstract tasks about frequencies difficult to interpret or to process.

In our previous discussion of these findings we tried to explain how the difficulty could arise (Evans *et al.*, 2003a). We suggested that people focus on the p mental model to begin with, but then try to divide this into the pq mental model and the p¬q mental model. The conditional probability, P(q|p), depends on the difference in probability between the probability of the pq mental model, P(pq), and the probability of p¬q mental model, P(p¬q ). If P(pq) is higher than P(p¬q), then P(q|p) is high. If P(pq) is lower than P(p¬q), then P(q|p) is low. When frequency information is given, as in the experiments on abstract conditionals, people will try to compare the number of pq instances with p¬q instances to make a judgement about P(pq) and P(p¬q). The difficulties of this may cause working memory problems, with the result that the response takes account of only the pq instances or too much account of these instances alone.

Another interesting possibility is suggested by the commentary of Edgington (2003). (See also Evans *et al.*, 2003a.) By a T3 account, judging an Adams conditional could lead to either the conditional or conjunctive probability response, depending upon how participants interpret the instructions. Suppose the statement was 'If the figure is a triangle then it is yellow'. According to the defective truth table, people would judge the truth value of this statement as follows:

| Yellow triangle | TT | True |
| Red triangle | TF | False |
| Yellow circle | FT | Irrelevant |
| Red circle | FF | Irrelevant |

Both Evans *et al.* (2003a) and Oberauer and Wilhelm (2003) asked their participants to judge the probability that the conditional was *true*. If the participants take the task to be that of judging the probability or believability of the conditional then they should come up with the conditional probability for the reasons we have explained. However, there is another interpretation of the question. Note that the Adams (T3) conditional is true in only one state of the world—TT or pq (Chapter 2). The probability that the world is in this state is P(pq), which equals the conjunctive probability. For example, if the question we asked participants was, 'What is the probability that a card drawn at random from the pack would make the statement true?', then this would be the correct answer for an Adams conditional. If Edgington is right, asking participants about the probability of a conditional, rather than its probability of truth, will produce significantly fewer conjunctive responses. And if she is right, participants of higher cognitive ability should give the conjunction response when the question is about probability of truth, for they will have the deeper grasp of when the conditional is true. If our original account is right, and the conjunctive response is caused by limited processing, then there will be no significant difference between asking about probability and probability of truth. Moreover, participants of lower cognitive ability will give the conjunctive response. Clearly, an empirical study is in order. (But see also Evans *et al.*, 2003a.)

Turning now to the mental representation of conditional statements, our theory is subtly and significantly different from the claims of Johnson-Laird and Byrne (1991, 2002) about the initial mental models that people will form of a conditional. We assume that the use of 'if' causes rapid and automatic focus on the p-possibility for all conditionals, similar to what Evans (1989) called the if-heuristic in an earlier theoretical framework (see Chapter 5). What happens next depends on whether or not people have background beliefs about the subject matter, or whether they are given frequency or other information relevant to the conditional.

Let us consider first the relatively abstract indicative conditionals, e.g. 'If the card is yellow then it has a circle printed on it.' People have little or no prior background beliefs about the subject matter of these conditionals. In an experiment, participants may, or may not, be given frequency information relevant to these conditionals. Suppose that participants in an experiment are given no frequency information, but are asked to assume that the conditional is true and to draw a conclusion that necessarily follows from it and a minor premise. In this case, people will suppose that there is no doubt about q given p. They will, initially at least, hypothetically set the degree of strength in q given p at the highest level. That is, people will assume that such conditionals are true, or at least extremely probable, unless provided with evidence to the contrary, which may come from other premises in 'suppression' experiments. Consistent with this hypothesis is the observation that for such conditionals MP rates are near 100% as are rates of identification of TT as a verifying case in truth table tasks (Chapter 3). However, there is a pragmatic effect even on such abstract conditionals. Affirmation of the Consequent (AC) rates tend to be very high for such conditionals in most studies (Chapter 3). Hence, we believe that without context, 'biconditional readings' are common and consist of a converse pragmatic inference of 'if q then p'.

As we pointed out earlier, in Chapter 6, the concept of biconditionality is problematic. In the case of a truth functional conditional, 'if p then q', the result of the converse implicature, 'if q then p', and of the inverse implicature, 'if not-p then not-q', are equivalent. The overall result would be to rule out the FT as well as the TF case. It is problematic then that in psychological experiments the two implicatures do not seem to result in an equivalence. For example, we have shown that in general AC inferences are more common with abstract conditionals and Denial of the Antecedent (DA) inference more common with thematic conditionals in most contexts (see Chapters 3 and 6). A strength of our own account is that the two are *not* equivalent. Adding the converse implicature, as commonly seems to occur with abstract conditionals, conjoins two statements 'if p then q' and 'if q then p', each of which has a defective truth table. Hence, while the former disallows p¬q and the latter ¬pq, the ¬p¬q or FF case remains irrelevant. This is a truth table pattern called 'defective biconditional' that we discussed in Chapter 3. It does not directly support the DA inference as a truth functional account would require. Similarly, while adding an inverse implicature to a realistic conditional supports DA and makes the FF (¬p¬q) case seem true rather than irrelevant (see Chapter 6), it does not immediately support the AC inference.

How can we represent this distinction in mental models for the conditional? Clearly not by the extensional (truth functional) approach of Johnson-Laird and Byrne. Whereas the JLB initial models of an abstract conditional are symbolized as:

pq
. . .

our own can be symbolized as:

$p \rightarrow q$ (0.95)
[$p \leftarrow q$ (0.80)]

This notation represents directionality, propensity, and implicature, all of which are absent in the JLB system. The arrows indicate the first two properties and can be read as conditional belief. So we can read the first line above as the tendency to believe q given p is 95%. The second line is an added, converse implicature (indicated by the square brackets) and can be read, the tendency to believe p given q is 80%. We believe that the fact that this is an implicature must be represented in the model, because under strict deductive reasoning instructions, people tend to suppress these implicatures, which occur by default from rapid System 1 processing (we say more on this later). Hence, they must represent implicatures in a way that distinguishes them.

There are some other important points to make about our symbols. People will only very rarely be able to report degrees of belief as precise as 0.95 or 0.80. They may be able to do this when they know relevant frequency distributions. More generally, however, people will only be able to say that their conditional degree of belief is relatively high or relatively low, or that it falls somewhere on a simple scale. The arrows represent such subjective degrees of conditional belief. The arrows do not refer to another, mental 'if' in a mental logic. That would be circular and have absurd consequences.

Consider the following thematic conditional about a coin we know to be fair:

If we spin the coin (s) then it will come up heads (h)

We would symbolize our mental state as:

s → h (0.50)

The above symbols mean that our conditional degree of belief that the coin will come up heads, given that it is spun, is 0.50. It does not mean that we have a conditional, in a mental logic, equivalent to 'if we spin the coin then our confidence it will come up heads is 0.50.' (See Edgington, 1995, p. 269, for a reductio ad absurdum of this supposed equivalence.) The arrow does not refer to any mental conditional, but rather to a disposition or propensity we have to believe and to assert h given s, including the disposition to give 0.50 as the judged probability of the thematic conditional. Thus the number 0.50 after the arrow measures the strength of this disposition or propensity. The arrow does not refer to an objective relation of any kind, such as a causal relation. Our conditional degrees of belief may be based on our beliefs about causal and other relations in the world, or on objective frequency distributions, but that is an epistemic and not a semantic fact. Finally, note that we have not, for this example, represented a degree of belief, acquired pragmatically, in our having spun the coin given that it comes up heads. Such a pragmatic inference from the statement of a realistic conditional would be highly sensitive to the context, e.g. whether anyone else was spinning coins in it.

We use our arrow symbol to refer to people's degrees of conditional belief. But how do people acquire these degrees of belief, however vague they are? To answer this question, the study of conditional reasoning must be integrated with judgement and decision making. There are many heuristics studied by researchers in judgement and decision making that can result in a judgement about whether $P(pq)$ is higher or lower than $P(p¬q)$. For example, a pq mental model might come much more readily to mind than a p¬q mental model, as a result of the availability heuristic (Tversky and Kahneman, 1973), leading to a high $P(q|p)$ judgement without an explicit comparison between the two mental models. For some people, p¬q could be the much more available mental model, because of their experience with disabling conditions or counter-examples for 'if p then q'. These people might judge $P(q|p)$ to be low without explicitly constructing the pq mental model. People can use a closeness or proximity heuristic (Teigen, 1998, 2004; Roese, 2004) to judge that one possibility is closer than another and so more probable given their background beliefs. For example, consider asserting about a precariously balanced teapot, 'If we are clumsy, it will fall.' We judge that our being clumsy and its falling is a much closer possibility than our being clumsy and its not falling, owing to the fact that it is close to falling anyway in the simplest sense. This judgement would then give us a high conditional degree belief that it will fall given that we are clumsy. We do not even have to judge in this case how probable it is that we will be clumsy (in spite of the claims of Edgington, 2003).

We pointed out in Chapter 2 that philosophers always seem to think of the Ramsey test as if it necessarily consisted of an explicit sequence of inferences. They imagine it to be, in effect, a pure System 2 process, in our dual process theory terms. As we discussed in Chapter 7, the 'heuristic' that is most like this view of the Ramsey test is the simulation heuristic (Kahneman and Tversky, 1982), which is a kind of 'heuristic' that for us is more in System 2 than System 1. The Ramsey test always requires some System 2 thought, and it can, but does not have to, call heavily on System 2 to derive a conditional degree of belief. It could compare pq and p¬q by running a simulation from p and aiming at q, or at ¬q, or at first one and then the other to try to avoid biases. It could compare pq and p¬q by constructing from p in System 2 a sequence of explicit inductive or probabilistic inferences of the kind studied by psychologists (for reviews and introductions to this literature, see Heit, 2000; Keren and Teigen, 2004; Shanks, 2004; Lagnado and Sloman, 2004). Here too it might be best to think of grounds for ¬q as well as for q, to try to overcome pitfalls such as the 'myside' bias (Baron, 2000). However, many System 1 processes, which are heuristics in our preferred sense of the term, could also be employed as aspects of a Ramsey test. As we have noted from the start, the Ramsey test is a high-level description of many processes that contribute to hypothetical thought. Describing fully the processes that can make up a Ramsey test of a conditional is a formidable challenge for psychological research on conditionals and in judgement and decision making.

When indicative conditionals are used in causal contexts (Over and Evans, 2003; Over *et al.*, in preparation), there is some tendency for a negative influence of $P(q|\neg p)$ on people's probability judgements about 'if p then q'. This result could mean that people take a causal conditional of this form to make the *semantic statement* that p causes q and realize that this logically implies that p raises the probability of q. For p to raise the probability of q, $P(q|p)$ has to be greater than $P(q|\neg p)$. As we explained in Chapters 7 and 8, the $\Delta p$ statistic, $P(q|p) - P(q|\neg p)$, measures the extent to which p raises the probability of q. People would be implicitly conforming to the $\Delta p$ statistic if they thought that causal conditionals semantically stated that this difference was high. The evidence (Chapter 8) is that people make limited use of $P(q|\neg p)$ and are influenced much less by this than $P(q|p)$.

This leads us to consider an alternative, pragmatic interpretation. Causal conditionals fall into the category where inverse pragmatic implicature is common. For example, they are associated with high rates of DA inferences and high ratings of FF cases as 'true' (Newstead *et al.*, 1997). It may be that a 'causal' conditional is simply a conditional that is epistemically based on, or justified by, a belief in a causal relation. Confidence in such a conditional might equal $P(q|p)$, but people would often infer pragmatically 'if not-p then not-q' from its assertion in a context that seemed to be about causation. Of course, $P(\neg q|\neg p)$ is high if and only if $P(q|\neg p)$ is low, and thus people's pragmatic inference here might reflect their knowledge that p should raise the probability of q if p causes q.

For example, consider:

9.1   If global warming continues then London will be flooded

This conditional might semantically state that the existence of a causal relation between global warming and the flooding of London. In that case, the conditional itself logically implies that global warming will raise the probability that London will be flooded. The other possibility is that 9.1 does not itself semantically make this statement, but speakers of it, in many contexts, would intend to convey pragmatically that global warming will raise the probability that London will be flooded. In this case, it would be pragmatically inappropriate, but not actually inconsistent, for speakers to assert 9.1 if they believed that London will be flooded in any case. On this view, the negative influence of $P(q|\neg p)$ reported in our experiments in Chapter 8 should be viewed as a positive influence of $P(\neg q|\neg p)$ and the relatively small size of the effect attributed to its origin in pragmatic inference. A conditional added pragmatically could be expected to be weaker in the participant's representation than one asserted directly.

It is important to answer the question of what a causal conditional asserts, as a result of its semantic content, and what can at times be inferred pragmatically from it. The answer would do much to settle the further question of whether T2 or T3 is the best account of 'causal' conditionals. Suppose 9.1 semantically states that there is a causal relation between global warming and the flooding of London. That makes 9.1 true even if its antecedent is false, i.e. global warming does not continue, and that implies T2 for 9.1 The causal relation is still a fact about the actual state of affairs whether the antecedent is true or not. Of course, this would in turn imply, by Lewis's proof (Chapter 2), that the probability of causal conditionals could not, in general, be the conditional probability. The same would be true of the counterfactuals that would be closely related to these conditionals (see Chapter 7). For instance, the probability of 9.1 would be the probability that the causal relation existed, and that might not equal the conditional probability, although it might be hard to separate the two in experiments (see Chapters 2 and 7).

We already have evidence that people evaluate the probability of causal conditionals such as 9.1 *primarily* in reference to the conditional probability (see Chapter 8). If the probability of these conditionals were precisely the conditional probability, that would imply T3 for them. But T2 could still hold for these conditionals, with their probability close to the conditional probability (see Chapter 2). In our view, even if T2 holds, the Ramsey test would still be part of the psychological evaluation of causal conditionals. Two Ramsey tests, for instance, would be called on to assess the $\Delta p$ statistic, $P(q|p) - P(q|\neg p)$. (Ramsey tests would also have to be part of other ways of assessing causal strength or power, as in Cheng, 1997.) Then too, the bounded nature of human thinking would imply some tendency to focus on one Ramsey test and $P(q|p)$, explaining the relatively weak influence of $P(q|\neg p)$ in our experiments on the probability of conditionals. This is a place where the psychology of conditionals should be linked with the psychology of bounded causal judgement, where it is argued that using conditional probability alone is often an efficient way to discover causal relations (Anderson, 1990; Shanks, 1995; Stanovich and West, 1998b, 2003; Over and Green, 2001).

## A suppositional account of the Wason selection task

Turning to other evidence for our view of the Ramsey test and the hypothetical evaluation of conditionals, we should consider next the Wason selection task, reviewed in Chapter 5. The standard abstract indicative selection task asks participants to test the truth of the conditional statement, as applied to four displayed cards. The participants have no background knowledge to help them assess this general conditional. They will try to evaluate the conditional by means of a Ramsey test, and in our symbols, we can represent their mental state as:

$$p \rightarrow q \, (?)$$

No degree of confidence is indicated in parentheses above. In a standard abstract task, the participants not only have no background knowledge, but are given no frequency information. Moreover, they are specifically told that the truth status of the conditional is unknown and that they are to try to discover it. With such impoverished content in the conditional, the Ramsey test reduces to the if-heuristic (Chapter 5) and the participants will focus on the antecedent possibility p. Such a default response seems to lead to a confirmation bias, just as the task's inventor, Peter Wason (1966), originally suggested. The participants in an abstract task should, of course, try to acquire a confidence in the general conditional by selecting the p and not-q cards. People of higher cognitive ability are more likely to select these cards (Stanovich and West, 1998). Higher cognitive ability will have that result by allowing these participants to represent both the pq and p¬q possibilities. But most participants will have some tendency, as a result of the if-heuristic, to focus too much on the p card.

On the standard abstract indicative (affirmative) selection task, the common choices are p alone or p and q. These choices would result from the conditional and biconditional interpretations that we have proposed above to account for findings of individual differences on the truth table and probability judgement tasks. The point here is that aiming to establish a directional mental link between p and q, $p \rightarrow q$, would only support the selection of the p card (expecting q). Only when there is also the goal of establishing a $p \leftarrow q$ link will people additionally choose q. Direct evidence for this is interpretation is provided in recent papers by Feeney and Handley, reviewed in Chapter 5 (Feeney and Handley, 2000; Handley *et al.*, 2002). In these experiments a second (logically irrelevant) conditional is presented to provide an alternative antecedent. This blocks q card selections in exactly the same way as similar manipulations in conditional inference studies block the AC inference (see Chapter 6). In both cases, we assume that the converse implicature is strongly inhibited.

What, however, of the evidence of matching bias on the selection task, reviewed in Chapter 5? We showed there that a quite simple account in terms of an if-heuristic and a matching-heuristic can account for the basic data. However, we suspect that the story is a bit more complicated. In particular, it may be that matching bias only operates when negative components are introduced and that something more like Wason's original verification bias account is correct when the conditional statement is affirmative, as suggested above. We suggest that when negated conditionals are used, people's mental representations are modified by pragmatic influences relating to the ordinary use of the word 'not'. In natural language we use negatives to deny presuppositions, thus directing attention to the affirmatives that are denied. Thus the matching cards will appear more relevant than the mismatching cards (Evans, 1989, 1995). In particular, there is a problem on the selection task in that the mismatching cards are implicit negations (7 must be taken as not 3 and so on). Evans *et al.* (1996a) showed that when explicit negations are used on the cards, the matching bias effect disappears. Conversely, people's tendency to draw conclusions in the conditional inference task is greatly reduced if they have to reason from a minor premise that implicitly negates one component of the conditional (Evans and Handley, 1999).

When the consequent is negated on the selection task, correct selections of the false-consequent card (now a matching card) increase dramatically (Chapter 5). With the statement 'If there is a B on one side of the card, then there is not a 6 on the other side of the card', most people will choose the logically correct B and 6 cards. What we think is happening here is that the Ramsey test is interacting with the pragmatics of 'not'. With the suppositional (but not truth functional) interpretation, 'if p then not q' is seen as the negation of 'if p then q'. This is because each conditional concerns only a hypothetical p possibility, and the latter statement affirms the existence of q given this possibility, while the former denies it. Direct evidence for this claim is available in a study by Pollard and Evans (1980). When participants were told that 'if p then q' was true, they judged that 'if p then not q' must be false and vice versa, with near universal frequency. Thus in the selection task, participants might well interpret the statement 'if p then not q' as an assertion that 'if p then q' is false. Hence, participants become focused on demonstrating falsity by looking for the p and q combination, choosing these cards. This is supported by protocol evidence (Wason and Evans, 1975; Evans, 1995). Verbal justifications for choosing p when the consequent is negative are typically given in terms of looking for a q to prove the statement false.

When the antecedent is negative, there is still a matching bias, but this combines with a strong tendency to choose also the true antecedent. So with a statement such as 'if there is a not a P on one side of the card, then there is a 1 on the other side of the card', there is a strong tendency to choose the letter that is not a P (say J), but also the P and the 1. We know that suppositional modelling of conditionals with negative antecedents will include the true-antecedent condition, not-p, as otherwise we could not account for the double negation effect that is so commonly observed in MT reasoning (If not p then q, not q, therefore not not p, therefore ?; see Chapter 3). We believe that people choose J and 1 here for the same reason as on the affirmative conditional. However, the pragmatic influence of 'not' makes the P card seem strongly relevant also, so people quite often select it as well.

As we explained in Chapters 5 and 8, we must distinguish between thematic, or realistic, selection tasks and abstract selection tasks. In an abstract task, participants are told that the conditional is only about the four cards displayed. This restriction is uncommon for thematic tasks, but far more important, of course, is that people come to these tasks with relevant background knowledge or experience. In some realistic indicative tasks, people will select the equivalent of the 'not-q' card, as we saw in Chapters 5 and 8. These are cases in which background knowledge, or a scenario given in the task, makes the p¬q possibility more highly available or salient or gives the participants a higher expectation that a p¬q instance could turn up. These cases are consistent with the hypothetical assessment of conditionals. To assess whether P(q|p) is high, people should compare the relative probabilities of pq and p¬q. Often there will be a tendency for people to focus on pq and to look for reasons why p and q might hold together. However, background beliefs, or the context in which the conditional is asserted, can give people a stronger or weaker tendency to focus more on p¬q and reasons why p and ¬q might hold together. A minority of high ability people will more often have a tendency to reflect on both pq and p¬q equally. Even a 'realistic' selection task is unrealistic in freely presenting the participants with cards having potential evidence on them. In the real world, it would be hopelessly inefficient to look for non-black things to test the claim that, if a bird is a raven, then it is black. In a 'realistic' selection task, it would be very easy to turn over a card with 'white' on it to find out whether 'raven' is on the other side. However, this is a point only those of higher cognitive ability are likely to grasp.

Deontic tasks are different again. What is relevant to the evaluation of a deontic conditional is not conditional probability alone, but conditional expected utility (Manktelow and Over, 1995; Over *et al.*, 2004). We would assert the deontic conditional, 'If we pick up the teapot then we should be careful', because we judged the expected benefits (or costs) of pq, i.e. our picking up the teapot and being careful, were higher (or lower) than p¬q, i.e. our picking up the teapot and not being careful. In short, we propose an expected utility version of the Ramsey test for the evaluation of deontic conditionals.

Not only is logical form different in deontic selection task, but the object or goal is different as well. The participants in an indicative selection task are asked to find out whether the deontic conditional is true or false. The participants in a deontic task are asked to find possible violations of the deontic conditional or rule, which is presupposed to be truly in force as a law or an agreement between people. Very often in deontic selection tasks, the pq possibility confers no benefit, while p¬q means a serious cost. For example, participants might be asked to imagine that they are pub owners who are checking on the law, 'If people are drinking alcohol then they must be at least 18 years old.' Pub owners get nothing by discovering an alcohol drinker who is 19, but they could suffer the serious longer-term cost of losing their licence by overlooking a 16 year old who is drinking alcohol. By discovering p¬q outcomes, i.e. underage drinking in our example, corrective action can be taken and the cost can be avoided. The result is that selecting the p and 'not-q' cards has the greatest expected utility, and most participants do choose these cards.

Our analysis also implies that authorities lay down a deontic conditional as a rule because of the expected benefits and costs of pq compared with p¬q. Health authorities in Britain got the drinking age rule adopted as a law because they judged the expected cost of drinking under the age of 18 to outweigh any 'benefit' there may be in that. Instead of yet more experimental research on deontic selection tasks, it would be better to have experimental studies of why deontic conditionals are asserted or accepted in the first place, i.e. before they get into deontic selection tasks (Over *et al.*, 2004). For this purpose, the decision theoretic analysis of deontic conditionals will have to be developed more fully, we would argue, and its relation to causal decision theory (Joyce, 1999) explored.

## Dual processes, pragmatics, and conditional reasoning

In this section we will discuss both pragmatic influences and the dual process theory of reasoning (see Chapter 1) as they are inextricably linked. What dual process accounts do is to describe the interplay of heuristic or pragmatic influences, in System 1, with abstract rule-based reasoning, in System 2. These systems can compete within individual participants in an experiment, the result being, for example, that a belief-based response can dominate in some individuals and a logic-based response in others.

There is overwhelming psychological evidence, reviewed in this book (see especially Chapters 5 and 6), that the way people understand conditionals and draw inferences from them is profoundly affected by the problem content and context, even when strict deductive reasoning instructions are employed in an experiment. The Ramsey test is a necessary first step in a psychological theory of the ordinary conditional, but there is much more to people's understanding and reasoning with conditionals. The Ramsey test as such does not tell us how people make conditional probability judgements, and it does not tell us about the pragmatics of people's use of conditionals. Even in accounting for people's treatment of abstract, indicative conditionals above, we had to discuss the pragmatic inference of 'if q then p' from a given 'if p then q'.

Let us start by summarizing what is known about the influence of prior knowledge and belief in deductive reasoning. Belief biases are known to operate both through conclusion and premise believability. All else being equal, people prefer to endorse conclusions that are believable rather than unbelievable, even though they are instructed to assess the logical validity of the argument. As we also saw in Chapter 6, people are also more willing to draw inferences from premises that they believe than ones that they do not believe. Research on belief bias provides some of the strongest evidence for dual process theory (Evans, 2003). People who have high measured general intelligence are better able to use System 2 to override System 1 and avoid biases. Not only do such people tend to find the normative solution more often on a range of reasoning and judgement problems, but they also are better able to resist pragmatic influences and belief bias when instructed to do so (Stanovich and West, 1997; Stanovich, 1999a). In old age, when System 2 to functioning is thought selectively to decline, people's ability to resist belief biases drop dramatically (Gilinsky and Judd, 1994). In addition, evidence from neural imaging studies shows that belief- and logic-based responses are associated with distinct brain areas (Goel and Dolan, 2003).

How is this relevant to conditional inference? Broadly, our interpretation of these effects is that System 1 inference, which includes many pragmatic inferences, is the default and that explicit deductive inference from assumptions occurs only when a special effort is made. People may make the special effort in ordinary circumstances when they are faced with a novel problem that cannot be solved by automatic System 1 processes. In most experiments on deductive reasoning, participants are explicitly asked, in effect, to make the special effort to use deductive inference from restricted assumptions. Some System 1 inferences are logically valid, but these inferences are from relevant beliefs and uncertain premises: they are not the result of explicit reasoning from restricted assumptions. System 1 inferences can implicitly comply with or conform to logical rules, but these inferences are not cases of explicitly following those rules.

System 1 inferences or *implicit* inferences do not always lead to logical error. What is characteristic of System 1 reasoning is that:

- inferences will be drawn rapidly and automatically without conscious reflection;
- prior knowledge and belief that is relevant in the context will be automatically added to premises;

- uncertainty in the premises will be taken into account, and some premises will be made uncertain by succeeding premises;

- inferences may be drawn on the basis of high probability or plausibility, rather than on the basis of what necessarily follows.

Implicit inferences—so defined—can lead to logically valid conclusions. Some trivially valid inferences are performed automatically by System 1 when people process the speech of another person. The speaker may, for example, say 'p & q', and the hearers may automatically infer q, especially if q is of special interest or relevance to them. Later the hearers may even incorrectly report that the speaker explicitly said q. The hearers might also report this in cases where the speaker only asserted 'if p then q', and the hearers believed strongly that p held. But System 1 can also infer 'if q then p' or 'if not-p then not-q' from 'if p then q'. These inferences are invalid, but can at times be pragmatically justified.

Implicit valid inferences can also be made in experiments in which the participants are given deductive reasoning instructions, i.e. they are told to assume the premises and draw only conclusions that necessarily follow from these restricted premises. These inferences are most likely to occur when participants are given abstract conditionals as premises, and they have no relevant prior beliefs about the subject matter. Examples reviewed in Chapter 2 include the drawing of MP, and the identification of the TT (pq) case as confirming the conditional with near 100% frequency. In neither case, in our view, is any deductive effort required. Similarly, studies of the Wason selection task have shown that the logically correct card choices of p on the indicative selection task and of both p and not-q cards on the deontic selection task (Chapter 5) can be achieved within the pragmatic mode of reasoning.

What evidence do we have for these claims? First, there is some direct evidence that MP as well as AC inferences are drawn implicitly during text comprehension (Rader and Sloutsky, 2002). However, we are arguing for implicit MP even in the kinds of task reviewed in this book, which are presented explicitly as reasoning problems. We have good reason to think that MP is normally made automatically by System 1 in such experiments, this apparently trivial inference being so easy to 'suppress' (see Chapter 6 for review of the relevant studies). A second conditional premise that suggests a disabling condition, i.e. a reason why p might not lead to q, will make endorsement rates of the conclusion of MP dramatically fail, as will use of conditionals that are known to bring such disabling conditions to mind based on rating studies with separate groups of participants. People will also resist the conclusion of MP if it appears unbelievable, or the premises are asserted by a source of low credibility or authority. Tips and warnings, where the speaker has no control over the consequent event, will lead to lower MP rates than promises and threats where the speaker does have control and so on. Suppression would not occur if people were using System 2 to make explicit valid inferences from premises that they assumed to be true and so unalterable.

In spite of these findings, it is clear that simple valid inferences such as MP *can* be made explicitly in System 2, as these belief biases and pragmatic influences can be greatly attenuated by putting extra stress on the deductive reasoning instructions (see Chapter 6). As already mentioned, the ability to override System 1, and to engage in deductive reasoning from assumptions held constant, is also related to general intelligence. This is why dual process theory must form an intrinsic part of any account of conditional reasoning.

We should also note again that people who take account of uncertainty in premises do not necessary think that the 'suppressed' inferences are invalid. They may only doubt that these inferences are *sound*, i.e. are valid *and* have true premises. It is rational by the highest Bayesian standards to take account of uncertainty in the premises of valid inferences. Where p is the conjunction of all the premises in a valid inference, and q the conclusion, $P(q|p)$ will of course be 1, but $P(q)$ should partly depend on $P(p)$. ($P(q)$ should depend on $P(q|not\text{-}p)$ and $P(not\text{-}p)$ as well.) How close people are to conforming to these standards is still an open question. But it is certainly not irrational to take account of $P(p)$ to make a judgement about $P(q)$ when $P(q|p)$ is 1 because there is a valid inference from p to q (Evans and Over, 1996, their Chapter 5).

Explicit MT inferences, from restricted assumptions, make use of reductio reasoning, much as the advocates of the mental logic theory have proposed (Rips, 1994; Braine and O'Brien, 1998a). That is to say, by supposing p one can derive q, which is contradicted by the information that not-q, leading to refutation of the supposition, and the conclusion not-p. This might seem a fairly natural process, but in our view, explicitly following the MT form does require explicit System 2 reasoning that goes beyond the automatic processing of the conditional statement. This is why rather less than 100% of adult participants (more like 70% in typical studies) endorse the valid conclusion, the rest saying that nothing follows. Moreover, there is strong evidence that adult participants, reasoning with abstract indicative conditionals, use supposition for MT.

First, there is the double negation effect discussed in Chapter 3. When the antecedent is negative, the argument to be evaluated has the structure: 'If not p then q; not q, therefore p'. People draw MT much less frequently in this case. The reductio reasoning will lead to the conclusion 'not-not-p', but people often fail to take the next valid step of inferring p. Second, there is the opposite conclusion effect (Evans, 1972b; Handley and Feeney, 2003; Schroyens *et al.*, 2003). Compare the following two problems, where the instruction is to evaluate the given conclusion by choosing one of the three options:

*Same conclusion*
> GIVEN
> If the letter is A then the number is 4
> The number is not 4
> CONCLUSION
> The letter is not A
> True   False   Can't tell

*Opposite conclusion*
> GIVEN
> If the letter is A then the number is 4
> The number is not 4
> CONCLUSION
> The letter is A
> True   False   Can't tell

In the 'same' conclusion problem, people should say, by means of MT, that the conclusion to be evaluated, that the letter is not A, is true. In 'opposite' conclusions, people should say, again by means of MT, the conclusion to be evaluated, that the letter is A, is false. The striking finding is that participants draw substantially more MT inferences when they are asked to evaluate the latter, 'opposite' conclusion. This indicates, in our view that people are reasoning suppositionally. The 'opposite' conclusion is what has to be assumed for MT and then refuted by reductio reasoning. This effect combines with the double negation effect. Hence, if the antecedents of the above arguments are negated, and the polarity of the presented conclusions reversed, opposite conclusions advantage is maintained, even though the overall rates are lower due to double negation (see Figure 9.1).

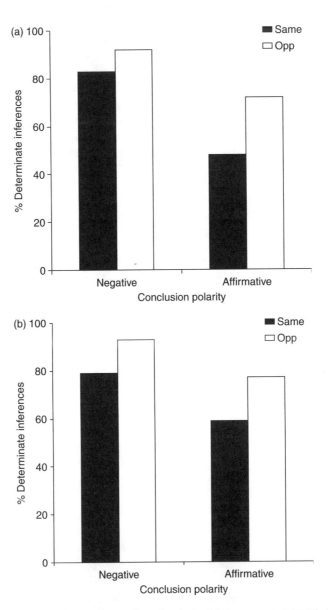

**Figure 9.1.** ModusTollens inference rates from the study of Schroyens *et al*. (2003). (a) Experiment 2 (*n* = 80). (b) Experiment 3 (*n* = 26).

While there is strong evidence that adults reasoning with abstract conditionals make MT using suppositional or reductio reasoning, this does not seem to be the case with younger children. Handley and Feeney (2003) reported a study of conditional reasoning administered to children of various ages as well as an adult group. They found that the 'opposite' conclusion advantage was not present in young children, but progressively increased through adolescence (see Figure 9.2). It was significantly present only in children of 15 years and older. However, substantial numbers of MT inferences are being drawn by younger children. In fact, many studies show higher rates of MT in adolescents than in adults (see Chapter 3). This strongly suggests that young children are inferring 'if not-q then not-p' pragmatically from 'if p then q', but that as they get older they increasing find this pragmatic inference unjustified.

The fallacious inferences, AC and DA, are not the result of a mental system entirely different from the system that produces MP and MT. It is not as though MP and MT are the result of only a System 2 mental logic, while AC and DA are purely pragmatic and the result of System 1. Usually AC and DA result from System 1 pragmatic inferences that, for example, automatically interpret a conditional assertion in some contexts as a biconditional. However, we do not always understand speakers automatically and immediately. Sometimes we have to reason explicitly in System 2 about what they might mean. Such reasoning could lead us to the conclusion that the speakers should be interpreted as inviting us to make a pragmatically justified AC or DA inference.

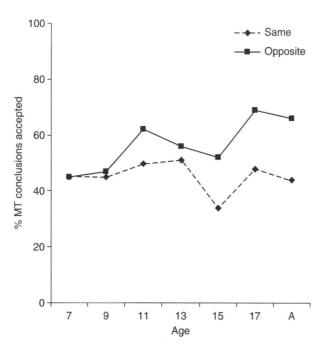

**Figure 9.2.** Modus Tollens rates for same and opposite conclusions at different ages. Data from Handley and Feeney (2003).

Evidence against the idea that valid inferences reflect one mental system and falla-cious inferences another is provided by the symmetry of the content and context effects that influences all four conditional inferences (Chapter 6). For example, the willingness to draw the valid MP and MT inferences is related to the perceived sufficiency of p for q, while the tendency to endorse DA and AC is related to the perceived necessity of p for q. In these cases, System 1 processes are automatically taking account of uncer-tainty in the premises. However, results are asymmetrical when strong deductive rea-soning instructions are used as in the study of Thompson (1994). These instructions stress that people are to restrict themselves to the given premises and to treat these as assumptions. Such instructions cause people to inhibit fallacious inferences relative to valid inferences. This result brings us back to dual process theory.

System 2 has the capacity, which is more effective in some people than others, to inhibit some pragmatic inferences and other automatic processes in System 1 (see, for example, Stanovich, 1999a). This fact about System 2 has clear links to the proposals of researchers working on executive processes in working memory (Gathercole, 2003). Neuropsychological studies of executive processes have pointed to a location in the right frontal lobe of the brain. We note in passing that recent research on reasoning using neural imaging techniques has found similar evidence, both for the inhibition of matching bias in conditional reasoning (Houde *et al.*, 2000), and for the inhibition of belief bias in syllogistic reasoning (Goel *et al.*, 2000; Goel and Dolan, 2003).

## Loose ends and future directions

We have qualified the title this chapter with the words 'towards' because we are more than aware of a number of issues to be addressed and problems to be solved in order to have a complete theory of conditionals. We identify some of the most burning of these issues in this section and try to indicate priorities for future research on 'if'.

The focus of this book has been on the understanding and use of conditionals, rather than on conditional reasoning *per se*. However, we have provided much discussion of research within the deduction paradigm, simply because that is what much of the psy-chological literature on conditionals has concerned itself with. We should note, then, that any complete theory of conditionals requires a theory of conditional inference. We have begun the task of devising such a theory as indicated in the previous section. We have not, however, referred directly to the problem of deductive competence. Although we believe that psychologists have overly emphasized logic and deductive reasoning instructions in their study of human reasoning (Evans, 2002a), there is nevertheless irre-ducible evidence that untrained adult participants have some logical competence (Evans and Over, 1996a, 1997). Moreover this competence develops with age and is more strongly apparent in adults with high general intelligence. It is therefore linked directly to System 2 functioning (Stanovich, 1999a).

One function that System 2 sometimes performs is to suppress pragmatic implica-tures arising in System 1, as we have discussed previously. However, another must be to provide effective algorithms for logical reasoning. On what is this based? Do we need to propose some form of mental logic of our own? Our view is that there is no innate mental logic in the sense of Rips (1994) or Braine and O'Brien (1991, 1998a). We see deductive reasoning as just one form of hypothetical thinking that System 2 can undertake. The minimal logical competence that people possess includes an understanding of the principle of non-contradiction, that both p and not-p are impossible. This principle is necessarily related to the ordinary linguistic meaning of 'not', which is denial. Non-contradiction is the basis of the reductio reasoning that results when inferences end in contradiction and a premise is negated, as in our earlier account of the System 2 thinking that underlies MT. If-elimination and if-introduction (for the ordinary conditional) are essential aspects of hypothetical thinking and the evaluation of 'if', just as the introduction and elimination rules for 'and' and 'or' are necessarily related to their meanings. We are doubtful, however, about the place of bivalence, that every statement is either true or false, and the implied the principle of excluded middle, that 'p or not-p' always holds, in the reasoning of ordinary people. For example, people have great difficulty with double negation elimination (see Chapter 3). There are also the truth value gaps that might be created by vagueness in natural language, and by the possible presence of an Adams, or T3, conditional.

In considering future research, we should note that it is one thing to propose that the conditional has a suppositional nature and another to have a well specified psychological account of the mental process involved. The Ramsey test does not tell us how people make conditional probability judgements but there is a wealth of other psychological research relevant to this. This work is on inductive inference of different types, heuristics, causal reasoning, and probability judgement in general. (For reviews and introductions to this literature, see Heit, 2000; Keren and Teigen, 2004; Shanks, 2004; Lagnado and Sloman, 2004.) However, there is almost no research at the moment that links conditional reasoning with inductive inferences or other mental processes that result in conditional probability judgements (but see Hadjichristidis et al., 2001). This gap must be filled, in our approach to the study of 'if', but that is beyond the scope of this book.

Another issue to be addressed psychologically is the problem of *closeness*. Just how do people conceive of some possibilities as closer than others to the actual state of the world or some given state of affairs? Most of the research on counterfactual thinking is to be found in the literatures on judgement and decision making and social psychology (Roese and Olson, 1995; Roese, 1997, 2004; Teigen, 1998, 2004). This research is largely ignored by psychologists who study conditional reasoning (but see Byrne and McEleney, 2000; McCloy and Byrne, 2000; Thompson and Byrne, 2002). It is again beyond the scope of this book to try to fill this gap, but that will have to be done before psychologists can get a deeper understanding of thematic conditionals, especially of 'causal' conditionals and counterfactuals (see Chapter 7, particularly the final sections). Of all the types of conditionals considered in this book, counterfactuals have received the least attention from those cognitive psychologists who specialize in the study of human reasoning. This is surprising, considering the depth of psychological interest that is intrinsic to the topic.

We have explained above how even the instructions in most deductive reasoning experiments limit what we can learn about conditionals from the results. Moreover, we are limited in our understanding of thematic conditionals until we obtain people's judgements about the probabilities of these conditionals and compare these with conditional probability judgements (but see Over and Evans, 2003, and Over et al., in preparation, will supply more information on this comparison.) We have ruled out a T1 account for the ordinary conditional, and the closely related extensional semantics of the JLB mental models theory. As we have also pointed out in Chapter 2, Lewis (1976) proved that a 'possible world' semantics for the ordinary conditional cannot be simply derived, as Stalnaker hoped, from the Ramsey test. By making these points, we have implicitly identified a void to be filled. What detailed semantic accounts are to be given of the various ordinary conditionals—indicative, counterfactual, deontic, and others— in natural language? All we can do here is give some indication of what our current knowledge suggests about the full semantic analysis of ordinary conditionals.

We have the best experimental evidence, at the present time, on the evaluation of indicative conditionals when frequency distributions are given, as in the experimental work reviewed in Chapter 8. Consider yet again:

9.2   If the card is yellow then it has a circle printed on it

This example is from the experiments of Evans *et al.* (2003a) and was discussed in Chapter 8. The participants were told that an example conditional such as 9.2 is about a random card drawn from a pack, consisting of red or yellow cards with circles or dia- monds printed on them. The frequency information, in our example, implies that there is a 1 in 5 chance that the card has a circle on it given that it is yellow. We found deci- sive evidence against the T1 hypothesis that participants will evaluate conditionals such as 9.2 as material, or truth functional, conditionals.

What of the T2 and T3 accounts? Suppose that 9.2 is a Stalnaker or other type of T2 conditional. Then 9.2 has a truth value in the FT and FF states of affairs, in which a yel- low card is not drawn. The truth value of 9.2 in FT and FF is determined by whether the TT possibility, in which a yellow circle is drawn, or the TF possibility, in which a yellow diamond is drawn, is closer to the FT and FF possibilities. But it is unclear how the participants could decide questions of closeness here, unless they used the relative frequency of the possibilities. In any case, we do not see how to elicit intuitive judge- ments of closeness from participants in experiments like this. That being so, we cannot make a prediction from T2 about the experiment. In contrast, T3 makes a clear predic- tion: the participants will respond with the conditional probability. If the suggestion of Edgington (2003), which we discussed above in this chapter, is correct, then T3 can even predict some conjunctive responses. Most participants respond with the condi- tional probability, and some with the conjunctive probability. These responses therefore appear to support a T3 account of conditionals such as 9.2 in examples like this.

Participants in our experiments on thematic or realistic conditionals such as 9.1 also respond with the conditional probability (Chapter 8; Over and Evans, 2003; Over *et al.*, in preparation). We even have an experiment showing similar trends for realistic coun- terfactuals (Evans *et al.*, in preparation, described in Chapter 8). A regression analysis shows that conditional probability is what predicts the responses. This again seems to be good support for T3. But Edgington's suggestion, from a T3 point of view, implies that conjunctive probability should also have a significant effect, and the regression analysis does not confirm this. On top of that, the probability of a Stalnaker conditional in these experiments might not differ very much from the conditional probability. Setting T2 and T3 decisively against each other will have to be done in further experiments.

It is possible to have tests of T2 because people will have intuitive judgements about closeness for many thematic conditionals such as 9.1. The same could be said about apparently related counterfactuals (Chapter 7), such as the statement that, if global warming had increased as rapidly in the nineteenth century as it is today, then London would already be flooded. People might well judge that a possibility in which global warming continues and London is flooded is closer than one in which global warming continues and London is not flooded. This judgement could result from vivid stories they have read in newspapers about global warming and flooding, or from a causal model in which global warming melts the polar icecaps. These judgements could lead to a probability judgement about 9.1, but it might be hard to find a significant difference between that and the conditional probability, as we have repeatedly pointed out above. Nevertheless, we would argue that separating T2 and T3 could come from experiments that investigate systematically, not only conditional probability judgements, but also closeness judgements.

As we have also pointed out above in this chapter, a T2 type account is implied if some conditionals, perhaps like 9.1, and closely related counterfactuals, semantically state the existence of a causal relation. But even in that case, the Ramsey test and hypothetical thought would be needed to evaluate conditionals in bounded thought, and as *part* of the assessment of causal strength or power, as we also pointed out above. Whether T2 or T3 is the better account of 'causal' conditionals and of counterfactuals is very much an open question. But the answer to it will not affect the importance of the Ramsey test and hypothetical thought in the psychological evaluation of these conditionals.

## Conclusions

In this book, we have tried to advance understanding of 'if' by considering together the arguments of philosophical logicians and the experimentation and theorizing of cognitive psychologists. We value each discipline equally but feel that the whole is greater than the sum of the parts. Psychological enquiry needs to be informed by philosophical arguments, and philosophical investigation is assisted by the empirical insights provided by the experimentalists. Combining the two requires hard discipline. We cannot satisfy ourselves, for example, with a logical analysis that lacks psychological plausibility or with a psychological account that ignores critical logical and philosophical arguments. In addition, psychology should not be divided against itself. We cannot explain conditional reasoning merely by modifying logical natural deduction rules and adding some pragmatics (Braine and O'Brien, 1991), nor by the mere 'modulation' of logical possibilities and adding some pragmatics (Johnson-Laird and Byrne, 2002). To make progress, the study of conditionals and conditional reasoning must be progressively and intimately integrated with research on judgement and decision making, on inductive and probabilistic inference, and on closeness judgements about possibilities.

We believe that we have made some progress. Natural language conditionals are not truth functional. The T1, material conditional hypothesis, is utterly incapable of accounting for a wide range of psychological data on how people understand and reasoning with conditional statements. 'If' must have, in some sense, a hypothetical or suppositional evaluation, even though there are difficult issues about its underlying semantics yet to be resolved. The varied and numerous uses of 'if' in everyday language that we have illustrated throughout this book all have something in common: the listener is invited to suppose or imagine a hypothetical state of affairs. This means that the apparent similarity of 'if' to propositional connectives such as 'or', 'and', and 'not' is an illusion. The ordinary conditional of natural language is not the 'if' of elementary textbooks on extensional logic.

The now extensive psychological literature on conditionals more than confirms the arguments of linguistics and logicians that readings of 'if' are many and varied according to the context in which it appears. The inferences that 'if' sanctions in the minds of the participants of psychological experiments depend upon the kind of use to which it is put (abstract or thematic indicative, counterfactual thought, inducement, advice, deontic rules, and others), but also on the prior knowledge and belief about particular contexts that participants bring with them to the laboratory. Inferences may be deductive, probabilistic, or pragmatic, with both implicit and explicit cognitive systems playing their part. The methodology of psychological experiments demands close scrutiny as well. People may make more or less effort to reason deductively or to take account of prior belief according to the precise instructions that they are given.

'If' is more than worthy of the academic effort devoted to its understanding since the ancient Greek philosophers first argued about it, and will doubtless remain a major focus of enquiry in cognitive science for many years to come. It is the linguistic device for stimulating hypothetical thought, and it is this facility—above all others—that identities the extraordinary and unique kind of intelligence that defines us as human beings.

# References

Adams, E. (1975). *The logic of conditionals: an application of probability to deductive logic*. Dordrecht: Reidel.

Adams, E. (1998). *A primer of probability logic*. Stanford: CLSI publications.

Adams, E. (1970). Subjunctive and indicative conditionals. *Foundations of Language, 6*, 89–94.

Ahn, W. and Graham, L. M. (1999). The impact of necessity and sufficiency information in the Wason four-card selection task. *Psychological Science, 10*, 237–242.

Allan, L. G. (1980). A note on the measurement of contingency between two binary variables in judgment tasks. *Bulletin of the Psychometric Society, 15*, 147–149.

Almor, A. (2003). Specialised behaviour without specialised modules. In D. E. Over (ed), *Evolution and the psychology of thinking* (pp. 101–120). Hove: Psychology Press.

Almor, A. and Sloman, S. A. (1996). Is deontic reasoning special? *Psychological Review, 103*, 374–378.

Anderson, J. R. (1990). *The adaptive character of thought*. Hillsdale, NJ: Erlbaum.

Andrews, A. D. (1993). Mental models and tableau logic. *Behavioral and Brain Sciences, 16*, 334.

Ball, L. J., Lucas, E. J., Miles, J. N. V., and Gale, A. G. (2003). Inspection times and the selection task: What do eye-movements reveal about relevance effects. *Quarterly Journal of Experimental Psychology, 56A*, 1053–1077.

Baron, J. (1994). Nonconsequentialist decisions. *Behavioral and Brain Sciences, 17*, 1–42.

Baron, J. (2000). *Thinking and deciding* (3rd edn). Cambridge: Cambridge University Press.

Barrouillet, P. and Lecas, J.-F. (1998). How can mental models theory account for content effects in conditional reasoning? A developmental perspective. *Cognition, 67*, 209–253.

Barrouillet, P. and Lecas, J.-F. (1999). Mental models in conditional reasoning and working memory. *Thinking and Reasoning, 5*, 289–302.

Barrouillet, P., Grosset, N., and Lecas, J.-F. (2000). Conditional reasoning by mental models: chronometric and developmental evidence. *Cognition, 75*, 237–266.

Barrouillet, P., Markovits, H., and Quinn, S. (2001). Developmental and content effects in reasoning with causal conditionals. *Journal of Experimental Child Psychology, 81*, 235–248.

Bennett, J. (2003). *A philosophical guide to conditionals*. Oxford: Oxford University Press.

Bonnefon, J.-B. and Hilton, D. J. (2002). The suppression of Modus Ponens as a case of pragmatic preconditional reasoning. *Thinking and Reasoning, 8*, 21–40.

Bradley, R. (2000). Conditionals and the logic of decision. *Philosophy of Science Proceedings, 67*, S18–S32.

Braine, M. D. S. (1978). On the relation between the natural logic of reasoning and standard logic. *Psychological Review, 85*, 1–21.

Braine, M. D. S. (1993). Mental models cannot exclude mental logic and make little sense without it. *Behavioral and Brain Sciences, 16*, 338–339.

Braine, M. D. S. and O'Brien, D. P. (1991). A theory of If: a lexical entry, reasoning program, and pragmatic principles. *Psychological Review, 98*, 182–203.

Braine, M. D. S. and O'Brien, D. P. (1998a). (eds) *Mental logic*. Mahwah, NJ: Lawrence Erlbaum Associates.

Braine, M. D. S. and O'Brien, D. P. (1998b). The theory of mental-propositional logic: description and illustration. In M. D. S. Braine and D. P. O'Brien (eds), *Mental logic* (pp. 79–89). Mahwah, NJ: Lawrence Erlbaum Associates.

Byrne, R. M. J. (1989). Suppressing valid inferences with conditionals. *Cognition, 31*, 61–83.

Byrne, R. M. J. (1991). Can valid inferences be suppressed? *Cognition, 39*, 71–78.

Byrne, R. M. J., Espino, O., and Santamaria, C. (1999). Counterexamples and the suppression of inferences. *Journal of Memory and Language, 40*, 347–373.

Byrne, R. M. J., and McEleney, A. (2000). Counterfactual thinking about actions and failures to act. *Journal of Experimental Psychology: Learning, Memory, and Cognition, 26*, 1318–1331.

Cheng, P. (1997). From covariation to causation: a causal power theory. *Psychological Review, 104*, 367–405.

Cheng, P. W. and Holyoak, K. J. (1985). Pragmatic reasoning schemas. *Cognitive Psychology, 17*, 391–416.

Cheng, P. W. and Holyoak, K. J. (1989). On the natural selection of reasoning theories. *Cognition, 33*, 285–314.

Chisolm, R. M. (1946). The contrary-to-fact conditional. *Mind, 55*, 289–307.

Cosmides, L. (1989). The logic of social exchange: has natural selection shaped how humans reason? *Cognition, 31*, 187–276.

Cosmides, L. and Tooby, J. (1992). Cognitive adapations for social exchange. In J. H. Barkow, L. Cosmides, and J. Tooby (eds), *The adapted mind: evolutionary psychology and the generation of culture* (pp. 163–228). Oxford: Oxford University Press.

Cosmides, L. and Tooby, J. (2000). Consider the source: the evolution of adaptations for decoupling and metarepresentation. In D. Sperber (ed.), *Metarepresentations* (pp. 53–115). Oxford: Oxford University Press.

Cummins, D. D., Lubart, T., Alksnis, O., and Rist, R. (1991). Conditional reasoning and causation. *Memory and Cognition, 19*, 274–282.

De Neys, W., Schaeken, W., and d'Ydewalle, G. (2003). Inference suppression and semantic memory retrieval: every counterexample counts. *Memory and Cognition, 31*, 581–595.

Dieussaert, K., Schaeken, W., and d'Ydewalle, G. (2002). The relative contribution of content and context factors on the interpretation of conditionals. *Experimental Psychology, 49*, 181–195.

Dudman, V. H. (1984). Conditional interpretations of if-sentences. *The Australian Journal of Linguistics, 4*, 143–204.

Dudman, V. W. (1988). Indicative and subjunctive. *Analysis, 48*, 113–122.

Edgington, D. (1995). On conditionals. *Mind, 104*, 235–329.

Edgington, D. (1997). Commentary. In M. Woods, *Conditionals* (pp. 95–137). Oxford: Oxford University Press.

Edgington, D. (2001). Conditionals. In E. N. Zalta (ed), *Stanford encyclopedia of philosophy* (http://plato.stanford/edu/entries/conditionals). Stanford: Stanford University.

Edgington, D. (2003). What if? Questions about conditionals. *Mind and Language, 18*, 380–401.

Elio, R. and Pelletier, F. J. (1997). Belief change as propositional update. *Cognitive Science, 21*, 419–460.

Evans, J. St. B. T. (1972a). Interpretation and matching bias in a reasoning task. *Quarterly Journal of Experimental Psychology, 24*, 193–199.

Evans, J. St. B. T. (1972b). Reasoning with negatives. *British Journal of Psychology, 63*, 213–219.

Evans, J. St. B. T. (1975). On interpreting reasoning data: a reply to Van Duyne. *Cognition, 3*, 387–390.

Evans, J. St. B. T. (1977). Linguistic factors in reasoning. *Quarterly Journal of Experimental Psychology, 29*, 297–306.

Evans, J. St. B. T. and Beck, M. A. (1981). Directionality and temporal factors in conditional reasoning. *Current Psychological Research, 1*, 111–120.

Evans, J. St. B. T. (1982). *The psychology of deductive reasoning*. London: Routledge.

Evans, J. St. B. T. (1983). Linguistic determinants of bias in conditional reasoning. *Quarterly Journal of Experimental Psychology, 35A*, 635–644.

Evans, J. St. B. T. (1984). Heuristic and analytic processes in reasoning. *British Journal of Psychology, 75*, 451–468.

Evans, J. St. B. T. (1989). *Bias in human reasoning: causes and consequences*. Brighton: Erlbaum.

Evans, J. St. B. T. (1993). The mental model theory of conditional reasoning: critical appraisal and revision. *Cognition, 48*, 1–20.

Evans, J. St. B. T. (1995). Relevance and reasoning. In S. E. Newstead and J. St. B. T. Evans (eds), *Perspectives on thinking and reasoning* (pp. 147–172). Hove: Erlbaum.

Evans, J. St. B. T. (1996). Deciding before you think: relevance and reasoning in the selection task. *British Journal of Psychology, 87*, 223–240.

Evans, J. St. B. T. (1998a). Inspection times, relevance and reasoning: a reply to Roberts. *Quarterly Journal of Experimental Psychology, 51A*, 811–814.

Evans, J. St. B. T. (1998b). Matching bias in conditional reasoning: do we understand it after 25 years? *Thinking and Reasoning, 4*, 45–82.

Evans, J. St. B. T. (2002a). Logic and human reasoning: an assessment of the deduction paradigm. *Psychological Bulletin, 128*, 978–996.

Evans, J. St. B. T. (2002b). Matching bias and set sizes: a discussion of Yama (2001). *Thinking and Reasoning, 8*, 153–163.

Evans, J. St. B. T. (2003). In two minds: dual process accounts of reasoning. *Trends in Cognitive Sciences, 7,* 454–459.

Evans, J. St. B. T. and Beck, M. A. (1981). Directionality and temporal factors in conditional reasoning. *Current Psychological Research, 1,* 111–120.

Evans, J. St. B. T. and Clibbens, J. (1995). Perspective shifts in the selection task: reasoning or relevance? *Thinking and Reasoning,* 315–323.

Evans, J. St. B. T. and Handley, S. J. (1999). The role of negation in conditional inference. *Quarterly Journal of Experimental Psychology, 52A,* 739–769.

Evans, J. St. B. T. and Lynch, J. S. (1973). Matching bias in the selection task. *British Journal of Psychology, 64,* 391–397.

Evans, J. St. B. T. and Newstead, S. E. (1977). Language and reasoning: a study of temporal factors. *Cognition, 5,* 265–283.

Evans, J. St. B. T. and Over, D. E. (1996a). *Rationality and reasoning.* Hove: Psychology Press.

Evans, J. St. B. T. and Over, D. E. (1996b). Rationality in the selection task: epistemic utility versus uncertainty reduction. *Psychological Review, 103,* 356–363.

Evans, J. St. B. T. and Over, D. E. (1997). Rationality in reasoning: the case of deductive competence. *Current Psychology of Cognition, 16,* 3–38.

Evans, J. St. B. T. and Twyman-Musgrove, J. (1998). Conditional reasoning with inducements and advice. *Cognition, 69,* B11–B16.

Evans, J. St. B. T. and Wason, P. C. (1976). Rationalisation in a reasoning task. *British Journal of Psychology, 63,* 205–212.

Evans, J. St. B. T., Barston, J. L., and Pollard, P. (1983). On the conflict between logic and belief in syllogistic reasoning. *Memory and Cognition, 11,* 295–306.

Evans, J. St. B. T., Newstead, S. E., and Byrne, R. M. J. (1993). *Human reasoning: the psychology of deduction.* Hove: Erlbaum.

Evans, J. St. B. T., Allen, J. L., Newstead, S. E., and Pollard, P. (1994). Debiasing by instruction: the case of belief bias. *European Journal of Cognitive Psychology, 6,* 263–285.

Evans, J. St. B. T., Clibbens, J., and Rood, B. (1995). Bias in conditional inference: implications for mental models and mental logic. *Quarterly Journal of Experimental Psychology, 48A,* 644–670.

Evans, J. St. B. T., Clibbens, J., and Rood, B. (1996a). The role of implicit and explicit negation in conditional reasoning bias. *Journal of Memory and Language, 35,* 392–409.

Evans, J. St. B. T., Ellis, C. E., and Newstead, S. E. (1996b). On the mental representation of conditional sentences. *Quarterly Journal of Experimental Psychology, 49A,* 1086–1114.

Evans, J. St. B. T., Handley, S. J., Harper, C., and Johnson-Laird, P. N. (1999a). Reasoning about necessity and possibility: a test of the mental model theory of deduction. *Journal of Experimental Psychology: Learning, Memory and Cognition, 25,* 1495–1513.

Evans, J. St. B. T., Handley, S. H., and Harper, C. (2001). Necessity, possibility and belief: a study of syllogistic reasoning. *Quarterly Journal of Experimental Psychology, 54A,* 935–958.

Evans, J. St. B. T., Handley, S. H., and Over, D. E. (2003a). Conditionals and conditional probability. *Journal of Experimental Psychology: Learning, Memory and Cognition, 29,* 321–355.

Evans, J. St. B. T., Legrenzi, P., and Girotto, V. (1999b). The influence of linguistic form on reasoning: the case of matching bias. *Quarterly Journal of Experimental Psychology, 52A,* 185–216.

Evans, J. St. B. T., Over, D. E., and Handley, S. H. (2003b). A theory of hypothetical thinking. In D. Hardman and L. Maachi (eds), *Thinking: psychological perspectives on reasoning, judgement and decision making* (pp. 3–22). Chichester: Wiley.

Evans, J. St. B. T., Over, D. E., and Handley, S. H. (2003c). Rethinking the model theory of conditionals. In W. Schaeken, A. Vandierendonck, W. Schroyens, and G. d'Ydewalle (eds), *The mental model theory of reasoning: refinements and extensions.* Hove: Psychology Press.

Feeney, A. and Handley, S. H. (2000). The suppression of q card selections: evidence for deductive inference in Wason's selection task. *Quarterly Journal of Experimental Psychology, 53A,* 1224–1243.

Fiddick, L., Cosmides, L., and Tooby, J. (2000). No interpretation without representation: the role of domain-specific representations and inferences in the Wason selection task. *Cognition, 77,* 1–79.

Fillenbaum, S. (1975). If: some uses. *Psychological Research, 37,* 245–260.

Fillenbaum, S. (1976). Inducements: on phrasing and logic of conditional promises, threats and warnings. *Psychological Research, 38,* 231–250.

Fischhoff, B. (1982). For those condemned to study the past: heuristics and biases in hindsight. In D.Kahneman, P. Slovic, and A. Tversky (eds), *Judgement under uncertainty: heuristics and biases* (pp. 335–351). Cambridge: Cambridge University Press.

Fodor, J. (1983). *The modularity of mind.* Scranton, PA: Crowell.

Frege, G. (1970). Begriffsschrift, a formula language, modelled on arithmetic, for pure thought. In J. van Heijnoort, *Frege and Gödel, two fundamental texts in mathematical logic* (original publication, 1879) (pp. 1–82). Cambridge, MA: Harvard University Press.

Gardenfors, P. (1998). *Knowledge in flux: modeling the dynamics of epistemic states.* Cambridge, MA: MIT Press.

Gathercole, S. (2003). *Short-term and working memory.* London: Taylor and Francis.

Gentner, D. and Stevens, A. L. (1983). *Mental models.* Hillsdale, NJ: Erlbaum.

Gentzen, G. (1969). Investigations concerning logical deduction. In *Collected papers of Gerhard Gentzen* (original publication, 1934) (pp. 68–131). Amsterdam: North Holland.

George, C. (1995). The endorsement of the premises: assumption based or belief-based reasoning. *British Journal of Psychology, 86,* 93–111.

Gibbard, A. (1981). Two recent theories of conditionals. In W. Harper, R. Stalnaker, and C. T. Pearce (eds), *Ifs* (pp. 211–247). Dordrecht: Reidel.

Gibbard, A. and Harper, W. L. (1978). Counterfactuals and two kinds of expected utility. In C. Hooker, J. Leach, and E. McClennan (eds), *Foundations and applications of decision theory* (pp. 126–162). Dordrecht: Reidel.

Gilinsky, A. S. and Judd, B. B. (1994). Working memory and bias in reasoning across the life-span. *Psychology and Aging, 9*, 356–371.

Gilovich, T., Griffin, D., and Kahneman, D. (2002). *Heuristics and biases: the psychology of intuitive judgement*. Cambridge: Cambridge University Press.

Goel, V. and Dolan, R. J. (2003). Explaining modulation of reasoning by belief. *Cognition, 87*, B11–B22.

Goel, V., Buchel, C., Rith, C., and Olan, J. (2000). Dissociation of mechanisms underlying syllogistic reasoning. *NeuroImage, 12*, 504–514.

Golding, E. (1981). *The effect of past experience on problem solving*. London: British Psychological Society.

Goldvarg, Y. and Johnson-Laird, P. N. (2000). Illusions in modal reasoning. *Memory and Cognition, 28*, 282–294.

Goodman, N. (1947). The problem of counterfactual conditionals. *Journal of Philosophy, 44*, 113–128.

Goodwin, R. Q. and Wason, P. C. (1972). Degrees of insight. *British Journal of Psychology, 63*, 205–212.

Green, D. W. (1995). Externalisation, counter-examples and the abstract selection task. *Quarterly Journal of Experimental Psychology, 48A*, 424–446.

Green, D. W., Over, D. E., and Pyne, R. (1997). Probability and choice in the selection task. *Thinking and Reasoning, 3*, 209–236.

Grice, P. (1975). Logic and conversation. In P. Cole and J. L. Morgan (eds), *Studies in syntax* (pp. 41–58), Vol. 3: Speech Acts. New York: Academic Press.

Grice, P. (1989). *Studies in the way of words*. Cambridge, MA: Harvard University Press.

Griggs, R. A. (1983). The role of problem content in the selection task and in the THOG problem. In J. St. B. T. Evans (ed), *Thinking and reasoning: psychological approaches* (pp. 16–43). London: Routledge.

Griggs, R. A. (1984). Memory cueing and instructional effects on Wason's selection task. *Current Psychological Research and Reviews, 3*, 3–10.

Griggs, R. A. and Cox, J. R. (1982). The elusive thematic materials effect in the Wason selection task. *British Journal of Psychology, 73*, 407–420.

Hadjichristidis, C., Stevenson, R. J., Over, D. E., Sloman, S. A., Evans, J. St. B. T., and Feeney, A. (2001). On the evaluation of *If p then q* conditionals. In *Proceedings of the 23rd Annual Meeting of the Cognitive Science Society*, Edinburgh.

Hájek, A. (2004). Probabilities of conditionals—revisited. *Journal of Philosophical Logic, 18*, 423–428.

Handley, S. H. and Feeney, A. (2003). Representation, pragmatics and process in model-based reasoning. In W. Schaeken, A. Vandierendonck, W. Schroyens, and G. d'Ydewalle (eds), *The mental model theory of reasoning: refinements and extensions*. Hove: Psychology Press.

Handley, S. J., Feeney, A., and Harper, C. (2002). Alternative antecedents, probabilities and the suppression of fallacies on Wason's selection task. *Quarterly Journal of Experimental Psychology, 55A*, 799–813.

Hawkins, S. A. and Hastie, R. (1990). Hindsight: biased judgements of past events after the outcomes are known. *Psychological Bulletin, 107*, 311–327.

Heit, E. (2000). Properties of inductive reasoning. *Psychonomic Bulletin and Review,* *7*, 569–592.

Hilton, D. J. (1995). The social context of reasoning: conversational inference and rational judgment. *Psychological Bulletin, 107*, 248–271.

Hilton, D. J., Jaspars, J. M. F., and Clarke, D. D. (1990). Pragmatic conditional reasoning: context and content effects on the interpretation of causal assertions. *Journal of Pragmatics, 14*, 791–812.

Hodges, W. (1993). The logical content of theories of deduction. *Behavioral and Brain Sciences, 16*, 353.

Holyoak, K. and Cheng, P. (1995). Pragmatic reasoning with a point of view. *Thinking and Reasoning, 1*, 289–314.

Houde, O., Zago, L., Mellet, E., Moutier, S., Pineau, A. *et al*. (2000). Shifting from the perceptual brain to the logical brain: the neural impact of cognitive inhibition training. *Journal of Cognitive Neuroscience, 12*, 721–728.

Hume, D. (1902). *Enquiries concerning the human understanding and concerning the principle of morals* (original publication 1777). Oxford: Oxford University Press.

Jackson, F. (1987). *Conditionals*. Oxford: Blackwell.

Jackson, S. L. and Griggs, R. A. (1990). The elusive pragmatic reasoning schemas effect. *Quarterly Journal of Experimental Psychology, 42A*, 353–374.

Jeffrey, R. C. (1967). *Formal logic: its scope and limits*. New York: McGraw-Hill.

Johnson-Laird, P. N. (1983). *Mental models*. Cambridge: Cambridge University Press.

Johnson-Laird, P. N. (1995). Inference and mental models. In S. E. Newstead and J. St. B. T. Evans (eds), *Perspectives in thinking and reasoning* (pp. 115–146). Hove: Erlbaum (UK) Taylor and Francis.

Johnson-Laird, P. N. and Byrne, R. M. J. (1991). *Deduction*. Hove: Erlbaum.

Johnson-Laird, P. N. and Byrne, R. M. J. (1992). Modal reasoning, models and Manktelow and Over. *Cognition, 43*, 173–182.

Johnson-Laird, P. N. and Byrne, R. M. J. (2002). Conditionals: a theory of meaning, pragmatics and inference. *Psychological Review, 109*, 646–678.

Johnson-Laird, P. N. and Savary, F. (1999). Illusory inferences: a novel class of erroneous deductions. *Cognition, 71*, 191–299.

Johnson-Laird, P. N. and Tagart, J. (1969). How implication is understood. *American Journal of Psyhology, 2*, 367–373.

Johnson-Laird, P. N., Legrenzi, P., and Legrenzi, M. S. (1972). Reasoning and a sense of reality. *British Journal of Psychology, 63*, 395–400.

Johnson-Laird, P. N., Legrenzi, P., Girotto, V., Legrenzi, M. S., and Caverni, J. P. (1999). Naive probability: a mental model theory of extensional reasoning. *Psychological Review, 106*, 62–88.

Joyce, J. M. (1999). *The foundations of causal decision making*. Cambridge: Cambridge University Press.

Kahneman, D. and Frederick, S. (2002). Representativeness revisited: attribute substitution in intuitive judgement. In T. Gilovich, D. Griffin, and D. Kahneman (eds), *Heuristics and biases: the psychology of intuitive judgement* (pp. 49–81). Cambridge: Cambridge University Press.

Kahneman, D. and Miller, D. (1986). Norm theory: comparing reality to its alternatives. *Psychological Review, 93*, 136–153.

Kahneman, D. and Tversky, A. (1982). The simulation heuristic. In A. Kahneman, P. Slovic, and A. Tversky (eds), *Judgment under uncertainty: heuristics and biases* (pp. 201–210). Cambridge: Cambridge University Press.

Kahneman, D. and Varey, C. A. (1990). Propensities and counterfactuals: the loser who almost won. *Journal of Personality and Social Psychology, 59*, 1101–1110.

Keren, G. and Teigen, K. H. (2004). Yet another look at the heuristics and biases approach. In N. Harvey and D. J. Koehler (eds), *Blackwell handbook on judgment and decision making*. Oxford: Blackwell.

Kirby, K. N. (1994). Probabilities and utilities of fictional outcomes in Wason's four card selection task. *Cognition, 51*, 1–28.

Klauer, K. C. (1999). On the normative justification for information gain in Wason's selection task. *Psychological Review, 106*, 215–222.

Klauer, K. C., Musch, J., and Naumer, B. (2000). On belief bias in syllogistic reasoning. *Psychological Review, 107*, 852–884.

Kneale, W. and Kneale, M. (1962). *The development of logic*. Oxford: Oxford University Press.

Kripke, S. A. (1980). *Naming and necessity*. Cambridge, MA: Cambridge University Press.

Lagnado, D. and Sloman, S. A. (2004). Probabilistic reasoning and the inside/outside distinction. In N. Harvey and D. J. Koehler (eds), *Blackwell handbook on judgment and decision making*. Oxford: Blackwell Publishing.

Laming, D. (1996). On the analysis of irrational data selection: a critique of Oaksford and Chater (1994). *Psychological Review, 103*, 364–373.

Levi, I. (1996). *For the sake of argument: Ramsey test conditionals, inductive inference and nonmonotonic reasoning*. Cambridge: Cambridge University Press.

Lewis, D. (1973). *Counterfactuals*. Oxford: Basil Blackwell.

Lewis, D. (1976). Probabilities of conditionals and conditional probabilties. *Philsophical Review, 95*, 581–589.

Lewis, D. (1981). Causal decision theory. *Australasian Journal of Philosophy, 59*, 5–30.

Lewis, D. (1986). *On the plurality of worlds*. Oxford: Blackwell.

Liu, I.-M., Lo, K.-C., and Wu, J.-T. (1996). A probabilistic interpretation of 'If-Then'. *The Quarterly Journal of Experimental Psychology, 49A*, 828–844.

Love, R. E. and Kessler, C. M. (1995). Focusing in Wason's selection task: content and instruction effects. *Thinking and Reasoning, 1*, 153–182.

Lowe, E. J. (2002). *A survey of metaphysics*. Oxford: Oxford University Press.

Lycan, W. G. (2001). *Real conditionals*. Oxford: Clarendon Press.

Manktelow, K. I. (1999). *Reasoning and thinking*. Hove: Psychology Press.

Manktelow, K. I. and Evans, J. St. B. T. (1979). Facilitation of reasoning by realism: effect or non-effect? *British Journal of Psychology, 70*, 477–488.

Manktelow, K. I. and Over, D. E. (1990). Deontic thought and the selection task. In K. J. Gilhooly, M. T. Keane, R. H. Logie, and G. Erdos (eds), *Lines of thinking*, Vol. 1 (pp. 153–164). Chichester: Wiley.

Manktelow, K. I. and Over, D. E. (1991). Social roles and utilities in reasoning with deontic conditionals. *Cognition, 39*, 85–105.

Manktelow, K. I. and Over, D. E. (1995). Deontic reasoning. In S. E. Newstead and J. St. B. T. Evans (eds), *Perspectives on thinking and reasoning* (pp. 91–114). Hove: Erlbaum.

Markovits, H. (1984). Awareness of the 'possible' as a mediator of formal thinking in conditional reasoning problems. *British Journal of Psychology, 75*, 367–376.

Markovits, H. (1986). Familiarity effects in conditional reasoning. *Journal of Educational Psychology, 78*, 492–494.

Markovits, H. (2002). A mental model analysis of young children's conditional reasoning with meaningful premises. *Thinking and Reasoning, 6*, 335–347.

Markovits, H. and Barrouillet, P. (2002). The development of conditional reasoning: a mental model account. *Developmental Review, 22*, 5–36.

Markovits, H., Fleury, M.-L., Quinn, S., and Venet, M. (1998). The development of conditional reasoning and the structure of semantic memory. *Child Development, 69*, 742–755.

McCloy, R. and Byrne, R. M. J. (2000). Counterfactual thinking about controllable events. *Memory and Cognition, 28*, 1071–1078.

McGee, V. (1989). Conditional probabilities and compounds of conditionals. *Philsophical Review, 98*, 485–542.

Medvec, V. H., Madley, S. F., and Gilovich, T. (1995). When less is more: counterfactual thinking and satisfaction among Olympic medalists. *Journal of Personality and Social Psychology, 69*, 603–610.

Mellor, D. H. (1993). How to believe a conditional. *Journal of Philosophy, 90*, 233–248.

Newstead, S. E. and Griggs, R. A. (1983). Drawing inferences from quantified statements: a study of the square of opposition. *Journal of Verbal Learning and Verbal Behavior, 22*, 535–546.

Newstead, S. E. and Griggs, R. A. (1984). Fuzzy quantifiers as an explanation of set inclusion performance. *Psychological Research, 46*, 377–388.

Newstead, S. E., Ellis, C., Evans, J. St. B. T., and Dennis, I. (1997). Conditional reasoning with realistic material. *Thinking and Reasoning, 3*, 49–76.

Newstead, S. E., Pollard, P., Evans, J. St. B. T., and Allen, J. L. (1992). The source of belief bias effects in syllogistic reasoning. *Cognition, 45*, 257–284.

Newstead, S. E., Handley, S. H., and Buck, E. (1999). Falsifying mental models: testing the predictions of theories of syllogistic reasoning. *Journal of Memory and Language, 27*, 344–354.

Nickerson, R. S. (1996). Hempel's paradox and the Wason selection task: logical and psychological puzzles of confirmation. *Thinking and Reasoning, 2*, 1–32.

Nisbett, R. E. and Wilson, T. D. (1977). Telling more than we can know: verbal reports on mental processes. *Psychological Review, 84*, 231–295.

Noveck, I. and O'Brien, D. P. (1996). To what extent do pragmatic reasoning schemas affect performance on Wason's selection task. *Quarterly Journal of Experimental Psychology, 49A*, 463–489.

O'Brien, D. P. and Bonatti, L. (1999). The semantics of logical connectives and mental logic. *Current Psychology of Cognition, 18*, 87–97.

Oaksford, M. (2002). Contrast classes and matching bias as explanations of the effects of negation on conditional reasoning. *Thinking and Reasoning, 8*, 135–151.

Oaksford, M. and Chater, N. (1991). Against logicist cognitive science. *Mind and Language, 6*, 1–38.

Oaksford, M. and Chater, N. (1994). A rational analysis of the selection task as optimal data selection. *Psychological Review, 101*, 608–631.

Oaksford, M. and Chater, N. (1995). Information gain explains relevance which explains the selection task. *Cognition, 57*, 97–108.

Oaksford, M. and Chater, N. (1996). Rational explanation of the selection task. *Psychological Review, 103*, 381–391.

Oaksford, M. and Chater, N. (1998). *Rationality in an uncertain world.* Hove: Psychology Press.

Oaksford, M. and Chater, N. (2003a). Computational levels and conditional inference: a reply to Schroyens and Schaeken (2003). *Journal of Experimental Psychology: Learning, Memory and Cognition, 29*, 150–156.

Oaksford, M. and Chater, N. (2003b). Conditional probability and the cognitive science of conditional reasoning. *Mind and Language, 18*, 359–379.

Oaksford, M. and Stenning, K. (1992). Reasoning with conditional containing negated constituents. *Journal of Experimental Psychology: Learning, Memory and Cognition, 18*, 835–854.

Oaksford, M., Chater, N., and Grainger, B. (1999). Probabilistic effects in data selection. *Thinking and Reasoning, 5*, 193–244.

Oaksford, M., Chater, N., and Larkin, J. (2000). Probabilities and polarity biases in conditional inference. *Journal of Experimental Psychology: Learning, Memory and Cognition, 26*, 883–889.

Oberauer, K. and Wilhelm, O. (2003). The meaning(s) of conditionals: conditional probabilities, mental models and personal utlities. *Journal of Experimental Psychology: Learning, Memory and Cognition, 29*, 680–693.

Oberauer, K., Wilhelm, O., and Diaz, R. R. (1999). Bayesian rationality for the Wason selection task? A test of optimal data selection theory. *Thinking and Reasoning, 5*, 115–144.

Ormerod, T. C. and Richardson, H. (2003). On the generation and evaluation of inferences from single premises. *Memory and Cognition, 31*, 467–478.

Over, D. E. (1993). Deduction and degrees of belief. *Behavioral and Brain Sciences, 16*, 361–362.

Over, D. E. (2003). From massive modularity to metarepresentation: the evolution of higher cognition. In D. E. Over (ed), *Evolution and the psychology of thinking: the debate* (pp. 121–144). Hove: Psychology Press.

Over, D. E. (2004a). Naïve probability and its model theory. In V. Girotto and P. N. Johnson-Laird (eds), *The shape of reason. Essays in honour of Paolo Legrenzi.* Hove: Psychology Press.

Over, D. E. (2004b). The psychology of conditionals. In K. I. Manktelow and M. C. Chung (eds), *Psychology of reasoning: theoretical and historical perspectives.* Hove: Psychology Press.

Over, D. E. and Evans, J. St. B. T. (1994). Hits and misses: Kirby on the selection task. *Cognition, 52*, 235–243.

Over, D. E. and Evans, J. St. B. T. (1999). The meaning of mental logic. *Current Psychology of Cognition, 18*, 99–104.

Over, D. E. and Evans, J. St. B. T. (2000). Rational distinctions and adaptations. *Behavioral and Brain Sciences, 23*, 693–694.

Over, D. E. and Evans, J. St. B. T. (2003). The probability of conditionals: the psychological evidence. *Mind and Language, 18*, 340–358.

Over, D. E. and Green, D. W. (2001). Contingency, causation and adaptive inference. *Psychological Review, 108*, 682–684.

Over, D. E., Hadjichristidis, C., Evans, J. St. B. T., Handley, S. H., and Sloman, S. A. (in preparation). *The probability of ordinary indicative conditionals.*

Over, D. E. and Manktelow, K. I. (1993). Rationality, utility and deontic reasoning. In K. I. Manktelow and D. E. Over (eds), *Rationality* (pp. 231–259). London: Routledge.

Over, D. E., Manktelow, K. I., and Hadjichristidis, C. (2004). Conditions for the acceptance of deontic conditionals. *Canadian Journal of Experimental Psychology, 58*, 96–105.

Paris, S. G. (1973). Comprehension of language connectives and propositional logical relationships. *Journal of Experimental Child Psychology, 16*, 278–291.

Pearl, J. (2000). *Causality: models, reasoning and inference.* Cambridge: Cambridge University Press.

Politzer, G. and Bourmand, G. (2002). Deductive reasoning from uncertain conditionals. *British Journal of Psychology, 93*, 345–381.

Politzer, G. and Braine, M. D. S. (1991). Responses to inconsistent premises cannot count as suppression of valid inferences. *Cognition, 38*, 103–108.

Pollard, P. (1982). Human reasoning: some possible effects of availability. *Cognition, 12*, 65–96.

Pollard, P. and Evans, J. St. B. T. (1980). The influence of logic on conditional reasoning performance. *Quarterly Journal of Experimental Psychology, 32*, 605–624.

Pollard, P. and Evans, J. St. B. T. (1981). The effect of prior beliefs in reasoning: an associational interpretation. *British Journal of Psychology, 72*, 73–82.

Pollard, P. and Evans, J. St. B. T. (1983). The effect of experimentally contrived experience on reasoning performance. *Psychological Research*, 287–301.

Pollard, P. and Evans, J. St. B. T. (1987). On the relationship between content and context effects in reasoning. *American Journal of Psychology, 100*, 41–60.

Quinn, S. and Markovits, H. (1998). Conditional reasoning, causality and the structure of semantic inference. *Cognition, 68*, B93–B101.

Quinn, S. and Markovits, H. (2002). Conditional reasoning with causal premises: evidence for a retrieval model. *Thinking and Reasoning, 8*, 179–192.

Rader, A. W. and Sloutsky, V. M. (2002). Processing of logically valid and logically invalid conditional inferences in discourse comprehension. *Journal of Experimental Psychology: Learning, Memory and Cognition, 28*, 59–68.

Ramsey, F. P. (1990). General propositions and causality (original publication, 1931). In D. H. Mellor (ed), *Philosophical papers* (pp. 145–163). Cambridge: Cambridge University Press.

Reber, A. S. (1993). *Implicit learning and tacit knowledge.* Oxford: Oxford University Press.

Rips, L. J. (1994). *The psychology of proof.* Cambridge, MA: MIT Press.

Rips, L. J. and Marcus, S. L. (1977). Suppositions and the analysis of conditional sentences. In M. A. Just and P. A. Carpenter (eds), *Cognitive processes in comprehension* (pp. 185–219). New York: Wiley.

Roberts, M. J. (1998). Inspection times and the selection task: are they relevant? *Quarterly Journal of Experimental Psychology, 51A*, 781–810.

Roberts, M. J. and Newton, E. J. (2002). Inspection times, the change task, and the rapid-response selection task. *Quarterly Journal of Experimental Psychology, 54A*, 1031–1048.

Roese, N. J. (1997). Counterfactual thinking. *Psychological Bulletin, 121*, 133–148.

Roese, N. J. (2004). Twisted pair: counterfactual thinking and hindsight bias. In N. Harvey and D. J. Koehler (eds), *Blackwell handbook on judgment and decision making*. Oxford: Blackwell.

Roese, N. J. and Olson, J. M. (1995). (eds) *What might have been: the social psychology of counterfactual thinking*. Mahwah, NJ: Lawrence Erlbaum Associates.

Sanford, D. H. (1989). *If p, then q: conditionals and the foundation of reasoning*. London: Routledge.

Schroyens, W. and Schaeken, W. (2003). A critique of Oaksford, Chater and Larkin's (2000) conditional probability model of conditional reasoning. *Journal of Experimental Psychology: Learning, Memory and Cognition, 29*, 140–149.

Schroyens, W., Schaeken, W., Fias, W., and d'Ydewalle, G. (2000). Heuristic and analytic processes in propositional reasoning with negative conditionals. *Journal of Experimental Psychology: Learning, Memory and Cognition, 26*, 1713–1734.

Schroyens, W., Schaeken, W., and d'Ydewalle, G. (2001). The processing of negations in conditional reasoning: a meta-analytic study in mental models and/or mental logic theory. *Thinking and Reasoning, 7*, 121–172.

Schroyens, W., Handley, S. H., Evans, J. St. B. T., and Schaeken, W. (2003). Hypothetical thinking strategies in evaluating conditional arguments: opposites do not attract. Unpublished manuscript: University of Leuven.

Shafir, E., Simenson, I., and Tversky, A. (1993). Reason-based choice. *Cognition, 49*, 11–36.

Shanks, D. R. (1995). Is human learning rational? *Quarterly Journal of Experimental Psychology, 48A*, 257–279.

Shanks, D. R. (2004). Judging covariation and causation. In N. Harvey and D. J. Koehler (eds), *Blackwell handbook on judgment and decision making*. Oxford: Blackwell.

Shwarz, N. and Vaughn, L. A. (2002). The availability heuristic revisited: ease of recall and content of recall as distinct sources. In T. Gilovich, D. Griffin, and D. Kahneman (eds), *Heuristics and biases: the psychology of intuitive judgment* (pp. 103–119). Cambridge: Cambridge University Press.

Sloman, S. A. (1996). The empirical case for two systems of reasoning. *Psychological Bulletin, 119*, 3–22.

Sloman, S. A. and Over, D. E. (2003). Probability judgement from the inside out. In D. E. Over (ed), *Evolution and the psychology of thinking* (pp. 145–170). Hove: Psychology Press.

Sperber, D. and Girotto, V. (2002). Use or misuse of the selection task? Rejoinder to Fiddick, Cosmides and Tooby. *Cognition, 85*, 277–290.

Sperber, D. and Wilson, D. (1986). *Relevance*, Oxford: Basil Blackwell.

Sperber, D. and Wilson, D. (1995). *Relevance* (2nd edn). Oxford: Basil Blackwell.

Sperber, D., Cara, F., and Girotto, V. (1995). Relevance theory explains the selection task. *Cognition, 57*, 31–95.

Stalnaker, R. (1968). A theory of conditionals. *American Philosophical Quarterly Monograph Series, 2,* 98–112.

Stalnaker, R. (1975). Indicative conditionals. *Philosophia, 5,* 269–286.

Stalnaker, R. (1976). Letter to van Fraassen. In W. Harper and C. Hooker (eds), *Foundations of probability theory, statistical inference, and statistical theories of science* (pp. 302–306). Dordrecht: Reidel.

Stalnaker, R. (1984). *Inquiry.* Cambridge, Mass.: MIT Press.

Stalnaker, R. and Jeffrey, R. (1994). Conditionals as random variables. In E. Eells and B. Skyrms (eds), *Probability and conditionals* (pp. 31–46). Cambridge: Cambridge University Press.

Stalnaker, R. and Thomason, R. (1970). A semantic analysis of conditional logic. *Theoria, 36,* 23–42.

Stanovich, K. E. (1999). *Who is rational? Studies of individual differences in reasoning.* Mahway, NJ: Lawrence Elrbaum Associates.

Stanovich, K. E. and West, R. F. (1997). Reasoning independently of prior belief and individual differences in actively open-minded thinking. *Journal of Educational Psychology, 89,* 342–357.

Stanovich, K. E. and West, R. F. (1998a). Cognitive ability and variation in selection task performance. *Thinking and Reasoning, 4,* 193–230.

Stanovich, K. E. and West, R. F. (1998b). Who uses base rates and P(D/¬H)? An analysis of individual differences. *Memory and Cognition, 28,* 161–179.

Stanovich, K. E. and West, R. F. (2000). Individual differences in reasoning: implications for the rationality debate. *Behavioral and Brain Sciences, 23,* 645–726.

Stanovich, K. E. and West, R. F. (2003). Evolutionary versus instrumental goals: how evolutionary psychology misconceives human rationality. In D. E. Over (ed), *Evolution and the psychology of thinking* (pp. 171–230). Hove: Psychology Press.

Stevenson, R. J. and Over, D. E. (1995). Deduction from uncertain premises. *The Quarterly Journal of Experimental Psychology, 48A,* 613–643.

Stevenson, R. J. and Over, D. E. (2001). Reasoning form uncertain premises: effects of expertise and conversational context. *Thinking and Reasoning, 7,* 367–390.

Strawson, P. F. (1952). *Introduction to logical theory.* London: Methuen.

Teigen, K. H. (1998). When the unreal is more likely than the real: post hoc probability judgements and counterfactual closeness. *Thinking and Reasoning, 4,* 147–177.

Teigen, K. H. (2004). The proximity heuristic in judgments of accident probabilities. Unpublished manuscript: University of Oslo.

Tetlock, P. E. (2002). Theory-driven reasoning about plausible pasts and probable futures in World politics. In T. Gilovich, D. Griffin, and D. Kahneman (eds), *Heuristics and biases: the psychology of intuitive judgement* (pp. 582–600). Cambridge: Cambridge University Press.

Thomason, R. (1970). A Fitch-style formulation of conditional logic. *Logique et Analyse, 13,* 397–412.

Thompson, V. A. (1994). Interpretational factors in conditional reasoning. *Memory and Cognition, 22,* 742–758.

Thompson, V. A. (2000). The task-specific nature of domain-general reasoning. *Cognition, 76,* 209–268.

Thompson, V. A. (2001). Reasoning from false premises: the role of soundness in making logical deductions. *Canadian Journal of Experimental Psychology, 50,* 315–319.

Thompson, V. A. and Byrne, R. M. J. (2002). Reasoning counterfactually: making inferences about things that didn't happen. *Journal of Experimental Psychology: Learning, Memory and Cognition, 28,* 1–18.

Thompson, V. A. and Mann, J. (1995). Perceived necessity explains the dissociation between logic and meaning: the case of 'only if'. *Journal of Experimental Psychology: Learning, Memory and Cognition, 21,* 1554–1567.

Tversky, A. and Kahneman, D. (1973). Availability: a heuristic for judging frequency and probability. *Cognitive Psychology, 5,* 207–232.

Tversky, A. and Kahneman, D. (1983). Extensional versus intuitive reasoning: the conjunction fallacy in probability judgment. *Psychological Review, 90,* 293–315.

Van Duyne, P. C. (1976). Necessity and contingency in reasoning. *Acta Psychologica, 40,* 85–101.

Van Fraassen (1976). Probabilities of conditionals. In W. L. Harper and C. A. Hooker (eds), *Foundations of probability theory, statistical inference and statistical theories of science* (pp. 261–300). Dordrecht: Reidel.

Wason, P. C. (1960). On the failure to eliminate hypotheses in a conceptual task. *Quarterly Journal of Experimental Psychology,* 12–40.

Wason, P. C. (1966). Reasoning. In B. M. Foss (ed), *New horizons in psychology I* (pp. 106–137). Harmandsworth: Penguin.

Wason, P. C. (1972). In real life negatives are false. *Negation, Logique et Analyse,* 57–38.

Wason, P. C. and Brooks, P. G. (1979). THOG: the anatomy of a problem. *Psychological Research, 41,* 79–90.

Wason, P. C. and Evans, J. St. B. T. (1975). Dual processes in reasoning? *Cognition, 3,* 141–154.

Wason, P. C. and Johnson-Laird, P. N. (1972). *Psychology of reasoning: structure and content.* London: Batsford.

Wason, P. C. and Shapiro, D. (1971). Natural and contrived experience in a reasoning problem. *Quarterly Journal of Experimental Psychology, 23,* 63–71.

Whitehead, A. N. and Russell, B. (1962). *Principia Mathematica* (original publication, 1910). Cambridge: Cambridge University Press.

Wilkins, M. C. (1928). The effect of changed material on the ability to do formal syllogistic reasoning. *Archives of Psychology, 16,* no. 102.

Wittgenstein, L. (1961). *Conditionals* (original publication, 1921). Oxford: Oxford University Press.

Wobcke, W. (2000). An information-based theory of conditionals. *Notre Dame Journal of Formal Logic, 41,* 95–141.

Woods, M. (1997). *Conditionals.* Oxford: Oxford University Press.

Yama, H. (2001). Matching versus optimal data selection in the Wason selection task. *Thinking and Reasoning, 7,* 295–311.

Yang, Y. and Johnson-Laird, P. N. (2000). Illusions in quantified reasoning: how to make the impossible seem possible and vice versa. *Memory and Cognition, 28,* 452–465.

# Index